Workshop Practice Manual

"Essential Exercises for Creativity"

JATINDER MADAN

and

PRINCE PAL SINGH

ISBN: 978-93-5891-534-1

Disclaimer

The book covers topics on workshop practices which are meant to be performed with the help of machines, equipment and tools following established procedures and guidelines. Many workshop practices involve application of heat, force, sharp tools and moving parts. Although the description of the various processes involved in workshop practice has been provided with utmost care, it may not be complete from all aspects. The procedures described for each practice provide a general description and is not claimed to be a complete stepwise procedure. Therefore, it is recommended that in the beginning the workshop practices are performed by the learners under the supervision of a trained person.

CONTENTS

ABOUT THE AUTHORS

Dr. Jatinder Madan has teaching and research experience in mechanical engineering with a focus on manufacturing engineering and allied topics. He received his Ph.D. in Mechanical Engineering from IIT Delhi, New Delhi. He was a visiting researcher at the System Integration Division of the National Institute of Standards and Technology (NIST), Gaithersburg, MD, USA, during August (2011- 2013) and May – June 2019. He traveled to the USA multiple times for participating in ASME's Manufacturing Science and Engineering Conference (MSEC), which is an international event of repute where scientists from across the world showcase their manufacturing focused research work.
Currently, Dr. Madan is serving as Professor and Head of the Mechanical Engineering department at the Chandigarh College of Engineering and Technology (Degree Wing), Chandigarh, India. Dr. Madan has been instrumental in developing the workshops at his college and has been an instructor for the Workshop Practice course to the undergraduate engineering students. He has published several research papers in reputed international journals and international conferences, supervised PhDs, Master's Thesis and completed funded research projects. He has research interests in Sustainable Design and Manufacturing, CAD/CAM, Design Automation, Design for Manufacturing and Assembly (DFMA), and Smart Manufacturing.

Dr. Prince Pal Singh is a mechanical engineer with specialization in manufacturing engineering. He did M. Tech. in Manufacturing Systems Engineering from Sant Longowal Institute of Engineering and Technology, Longowal, Punjab, India and B. Tech. in Mechanical Engineering from Kurukshetra University, Kurukshetra, Haryana, India. He received Ph.D. in Mechanical Engineering with specialization in reconfigurable manufacturing systems from I.K. Gujral Punjab Technical University, Kapurthala, Punjab, India. He has taught several courses to undergraduate students of mechanical engineering, such as manufacturing systems, workshop technology and manufacturing processes. His current research focus is on manufacturing systems design, sustainable manufacturing, and reconfigurable manufacturing systems. He has published many research papers in reputed refereed journals and international conferences. He currently serves as Data Analyst in the Engineering and Analytics Department of the Spyglass Analytics Software Private Limited, Gurugram, Haryana, India.
The authors have contributed equally for preparing the manuscript.

FOREWORD

I am pleased to write a foreword for the Workshop Practice Manual co-authored by Dr. Jatinder Madan and Dr. Prince Pal Singh. The course on workshop practice is part of first year curriculum of Engineering Degree Programs in most of the Universities. This course provides much needed exposure to the students to exercise creativity, which leads to innovation and development at later stage.

The book covers the essential topics of basic workshop practices with a chapter on 3D Printing. Understanding of the conventional workshop practices is important to correctly understand the newer manufacturing processes. Further, 3D printing process has come as a disruptive technology with the potential of replacing the subtractive manufacturing techniques, presently in use. With the focus being on manufacturing processes, the book also covers power tools. In addition, authors have covered the safety precautions in the workshops, which is also an important aspect for learning often neglected. The book chapters are well designed and presented using several useful illustrations to explain the processes, tools, and exercises in a unique way. The response sheets given for each chapter make the learning experience very interactive.

Dr. Jatinder Madan and Dr. Prince Pal Singh have shared their rich experience in the field of Mechanical Engineering and Manufacturing Processes in this book. I am sure that the book will become a valuable resource in the hands of the learners and instructors of the workshop practice.

Dr. B S Pabla
Professor, Mechanical Engineering Department
National Institute of Technical Teachers Training and Research Chandigarh
Ex Principal, Chandigarh College of Engineering and Technology Chandigarh

PREFACE

This book is about learning workshop practices, techniques, and tools which are essential for exercising creativity, project work and making jobs. Workshop practice is generally introduced in undergraduate engineering programs to the freshman in several universities throughout the world. Several diploma-level and vocational courses also introduce workshop practice in their programs. The objective of this course is to provide the learners and students a practice-based exposure to the most basic manufacturing processes which we also call as workshop practices, such as carpentry, fitting, foundry (or casting), welding, sheet metal, electrical, electronics, smithy and forging, and machining. Although there have been notable developments in manufacturing technologies, with many new manufacturing processes being introduced, the importance of the above-mentioned workshop practices has not waned. Moreover, knowledge of the conventional processes is needed to correctly understand the newer manufacturing processes. In recent years, newer manufacturing processes like additive manufacturing or 3D printing have gained importance, which also needs to be introduced to the students in their first year so that they can use it for exercising creativity in the remaining years of their program. Power tools, such as hand drills are also very important to perform different tasks in many projects.

We got the idea of writing this book when we had to struggle a lot to make a customized manual for the workshop practice course for the undergraduate students. Most of the available resources and books provide a detailed discussion on the topics but a limited exposure to the practice-based learning. Moreover, safety, which is the most important aspect of learning in workshops, was described in the books on workshop practices to a limited extent only.

The topics in the book were selected to include the basic manufacturing processes, carpentry, fitting, foundry (or casting), welding, sheet metal, electrical, electronics, smithy and forging, and machining. Further, 3D printing processes, which have gained importance in recent times, have also been introduced. A chapter has been dedicated for describing essential power tools, which are often used for performing a variety of tasks and project work.

The book provides well-structured content for the learners to perform exercises on important workshop practices. Selected jobs are given at the end of each chapter along with the detailed procedure with illustrations. Response sheets have been provided at the end of each chapter where the learners can provide answers to specific questions after they make the given job to get feedback from the instructors.

Safety, which is the most important for carrying out workshop practices, has also been given due coverage throughout the book. Illustrations have been used to the maximum possible for clarity and understanding, making the learning of workshop practices enjoyable. We are hoping that the book will greatly help the students and learners of workshop practice.

We shall gratefully acknowledge the suggestions for the improvement of the book at Jatinder.madan@gmail.com or princepal123@gmail.com.

Jatinder Madan   Prince Pal Singh

ACKNOWLEDGMENTS

Initially, the book project on Workshop Practice Manual appeared to be simple. However, as we progressed, challenges of how to best describe the manufacturing processes/workshop practices and the exercises to be performed became apparent. We would like to thank Mr. Nishant Kumar and Mr. Vinay Dev Dutt, both senior (2019-23 batch) students of B. E. Mechanical Engineering program at Chandigarh College of Engineering and Technology (Degree Wing), Chandigarh, who helped in preparing the illustrations.

The support provided by the workshop staff of the Chandigarh College of Engineering and Technology (Degree Wing); Chandigarh was crucial during preparation of the manuscript. Mr. Manoj Kumar, Workshop Assistant, Mr. Malkeet Singh, Workshop Assistant and Mr. Mohit Kumar, Helper, whole heartedly supported us on various aspects of different chapters and refining the practical exercises.

No book project can be accomplished without support of the family. We are highly indebted to our families who supported us in this long project.

Authors

Chapter 1

INTRODUCTION AND SAFETY

1.1 INTRODUCTION TO WORKSHOP PRACTICES

Workshop is a place to acquire basic knowledge for making products ranging from simple models to complicated ones using different raw materials. Workshop practice helps in developing and enhancing hands on use of engineering materials, tools, equipment, processes, and machines as well as manufacturing techniques. Hence, to develop engineering skills, one must practice the basics of manufacturing processes by learning from the experience of craftsmen through practical demonstrations and practice in the workshop. The term manufacturing process refers to the method of processing raw materials to obtain various products in different shapes and sizes to accomplish functional requirements. Manufacturing processes may be classified into the following five groups.

- i. Primary shaping processes
- ii. Machining processes
- iii. Joining processes
- iv. Surface finish processes
- v. Processes affecting change in properties

1.1.1 Primary shaping processes

Primary shaping processes are those which help to make near net shape parts that are very close to the desired geometry. This means that requirement of additional finishing operations to obtain part geometry as per the required specifications is minimal. Some processes produce finished parts that may not require further work to finish products to the desired shape and size. For example, in the foundry shop, molten metal is poured into a cavity, which is also called mold. After the molten metal cools and solidifies, a near net shape part is produced. Such a part can either be used after minimal further processing, such as cleaning or can be machined to get an excellent finish. Other important primary shaping processes are injection molding, diecasting, smithy, and forging, bending, and rolling.

1.1.2 Machining processes

Parts that need further processing after primary processes to make it of the required shape and size by removing material. A typical machine shop has a number of machines, such as lathe, milling, drilling and shaper, which can help remove the material from the workpiece. Some of the common machining operations are turning, facing, threading, knurling, boring and drilling.

1.1.3 Joining processes

Joining processes are used in general fabrication work to join metal parts permanently or temporarily with the application of heat and pressure. Some of the joining processes are gas welding, arc welding, spot welding, soldering, and brazing.

1.1.4 Surface finish processes

The surface finishing processes are meant to improve the appearance of the parts that may be produced using any of the above-mentioned manufacturing processes. Sometimes a protective coating onto the metal may be added, or a negligible amount of metal from the component is removed to get a good surface finish. Some of the surface finish processes are buffing, grinding, and painting.

1.1.5 Processes affecting change in properties

Most of the engineering components are made up of metals with sufficient strength. However, certain engineering applications require better properties, such as good surface hardness. Therefore, heat treatment processes are required to enhance the properties, such as strength and wear resistance of these metallic components. The heat treatment processes generally involve heating the metallic component to a certain temperature and cooling it in a controlled environment to enhance its engineering properties. Examples of heat treatment processes are hardening, case hardening and normalizing. Other processes that change the material properties are cold working, hot working and shot peening.

1.2 WORKSHOP PRACTICES

A number of workshop practices are available, the most notable being carpentry, welding, machining, smithy, sheet metal working, foundry, fitting, and electrical technology. The above-mentioned workshop practices form part of the curriculum in most of the undergraduate engineering programs in the first year. Further, with the development of technology, 3D printing (also called additive manufacturing) has become very popular. Therefore, 3D printing has also been introduced in this book as a workshop practice. The manufacturing processes mentioned above are practiced in various shops. The above-mentioned practices are performed in respective shops, where required tools, equipment and machinery meant for carrying out different processes are available. A brief of the above-mentioned workshop practices is mentioned in the following paragraphs, details of which are discussed in the subsequent chapters. Figure 1.1 provides snapshots of a few products made using the above-mentioned workshop practices.

i.   *Carpentry shop*: This shop is used for making wooden products, such as door frames, window frames, and furniture.

ii.   *Foundry shop*: This shop is used for making parts by first melting the metal followed by pouring it into a mold cavity to get desired shaped products. Examples of products made in this shop are dumbbells and gears.

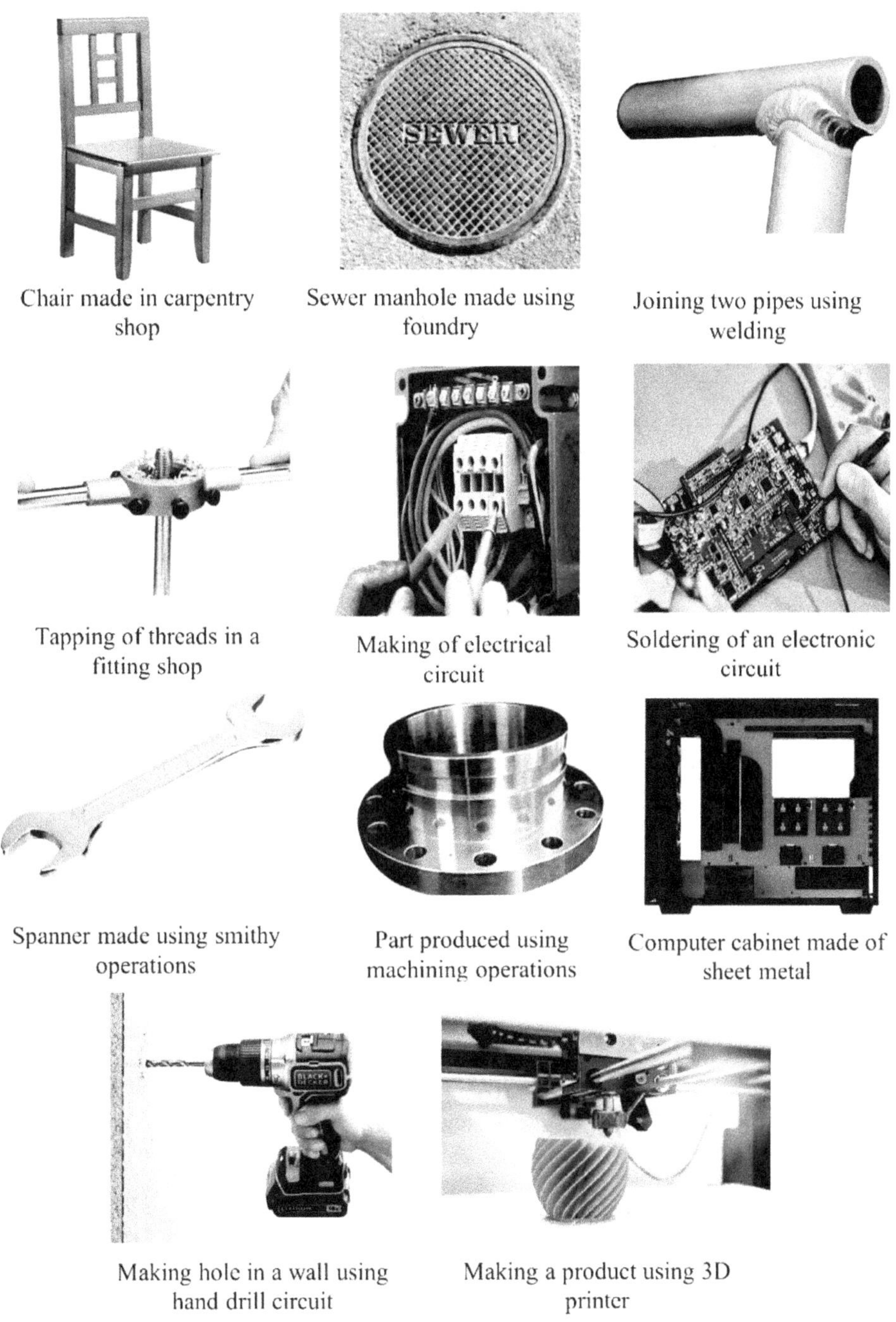

Chair made in carpentry shop

Sewer manhole made using foundry

Joining two pipes using welding

Tapping of threads in a fitting shop

Making of electrical circuit

Soldering of an electronic circuit

Spanner made using smithy operations

Part produced using machining operations

Computer cabinet made of sheet metal

Making hole in a wall using hand drill circuit

Making a product using 3D printer

Figure 1.1 Instances of applications of workshop practices

iii. *Welding shop:* This shop is used to join metallic pieces by obtaining coalescence on their edges with the applications of a welding process, such as arc welding, gas welding and spot welding.

iv. *Fitting shop:* Fitting operations are used to provide fitting of metallic jobs and components using hand tools, such as hacksaws and files.

v. *Electrical shop:* This shop is used to perform various processes related to electrical fitting, such as domestic wiring, fitting, and switchboard assembly.

vi. *Electronics shop:* This shop is used for the fabrication of products that require electronic components. Further, soldering of wires, which is commonly used to build circuit boards, is also practiced in this shop.

vii. *Black smithy shop:* This shop is used for heating the metal to the plastic stage and then applying force to shape it for making useful products, such as a chisel.

viii. *Machine shop:* This shop is used for shaping products by removing material from the workpiece using a variety of machines and cutting tools.

ix. *Sheet metal shop:* This shop is used to make products with thin metal sheets as raw material using hand tools. Examples of products made of sheet metal are funnel and tray.

x. *Power tools:* Several tools cannot be assigned to a particular shop but are helpful to carry out operations in different shops. Examples of such tools are power hacksaws, hand drills, bench drills and riveting guns. These power tools are also helpful for general projects and home improvement works.

xi. *3D printing:* This is a relatively new manufacturing process for which making CAD model or 3D model of the part to be made is an essential requirement. 3D printing provides ample freedom to designer to design products of any shape that otherwise cannot be made using traditional workshop practices.

1.3 GENERAL SAFETY PRECAUTIONS

A number of safety precautions are to be followed while doing workshop practices so that the operations are accomplished safely without any accidents. Safety is very important for workshop practices due to the reason that several tools are sharp and require application of force. Furthermore, shops like welding, smithy and foundry need working of the metal at very high temperatures. Therefore, it is important that the workshop practice is done under the supervision of a trained instructor and with utmost care following all standard operating procedures and safety precautions. This needs to be mentioned that not following the safety precautions carries the risk of accidents and injuries. In the following paragraphs, we discuss important safety precautions, which are generic and common to all the workshops.

1.3.1 Things to be avoided

i. Walk carefully and watch your steps in the workshop, don't run.

ii. Don't mess around in the workshop.

iii. Never remove or disable safety devices attached to an equipment.

iv. Never ignore danger signs which are displayed.

v. Don't use equipment that you think is unsafe.

vi. Don't use machinery until you've been trained to do so, or a competent person is supervising you.

vii. Don't operate machinery if there's someone within the operating safety area.
viii. Don't leave working machinery or equipment unattended.
ix. Ensure no distractions while you're working.
x. It's best practice not to drink, eat or smoke in a workshop area.
xi. When the machine is in motion, never leave it unattended.
xii. Never wear loose clothing, chains, rings, or watches at work.
xiii. Never use compressed air for cleaning a machine as it may blow in your face or eye and cause injury.

1.3.2 Things that must be done

Before working

i. Get the basic knowledge of machines, like knowing how to start and stop a machine.
ii. Make sure the machine you are going to work on is properly gripped or clamped.
iii. Before beginning any work, pay close attention to the instructor's directions
iv. You must know where the emergency STOP buttons, and isolating switches are.
v. Pay attention to the safety hazard signs.
vi. Put on the personal protective clothing and equipment (PPE), such as goggles, gloves, safety shoes and helmets.
vii. Your hair, clothes, and jewelry you wear should not be loose as these may get in your way and pose danger to your safety.
viii. Before connecting or disconnecting any electrical equipment, turn it off or isolate it.
ix. Always know about the first aid arrangements for your workshop.

While working

i. Concentrate on what you're doing.
ii. Follow instructions and procedures. Ask for guidance from the instructor in case of a doubt.
iii. Always keep your work area tidy and free from metal scrap, burrs, and unwanted components.
iv. Keep the working area and floor clean from oil, dirt, and grease.
v. Ensure that equipment doesn't extend out from the workbench further than necessary.
vi. Take care of inflammable substances in the workshop by keeping them in a designated place.
vii. Use the right tools for the job and concentrate on what you are doing.
viii. Keep the tools back in their designated places when not in use.
ix. Use machine guards provided.
x. Never clean the machine while it is in use; instead, disconnect it from the power source first.
xi. Report any faults in tools or equipment which could cause an accident.
xii. Be careful when using sharp tools or objects.
xiii. Keep to gangways; shortcuts are not to be used.
xiv. Follow safety signs and warning notices in the workshop.

1.3.3 Safety rules display

It is well known that *a picture is worth a thousand words*. Therefore, many workshops display safety precautions to be followed in the form of illustrative posters. Few illustrative posters are given in Figure 1.2 to create an impression in a learner's mind going for the workshop practice. It is strongly desirable that the learners carefully look at these posters and signboards and discuss them with their instructor if required to get more clarity.

Figure 1.2 Illustrations of safety practices in workshops

1.3.4 Hazards of violating safety

It is extremely important to know about the probable hazards if the safety precautions outlined in the previous section are ignored by the learners. The illustrations shown in Figure 1.3 and Figure 1.4 highlight few such hazards.

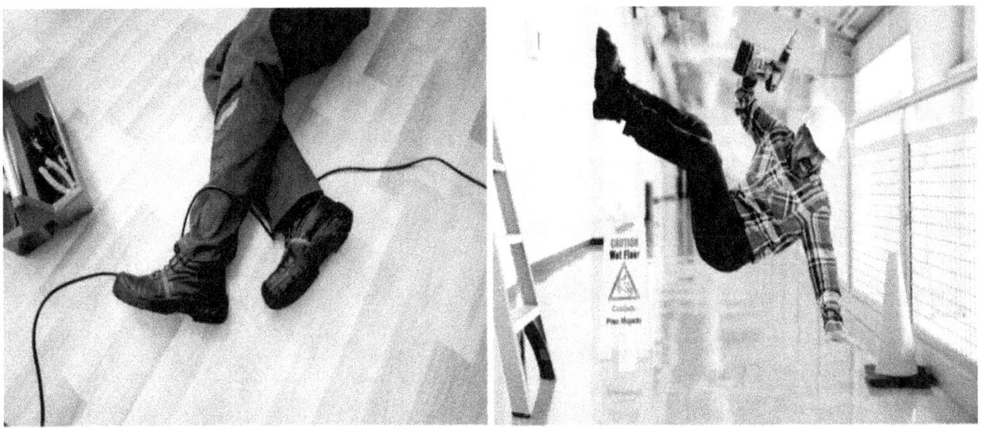

Figure 1.3 Illustrations showing the dangers of untidy workplace and oil spillage or wet floor

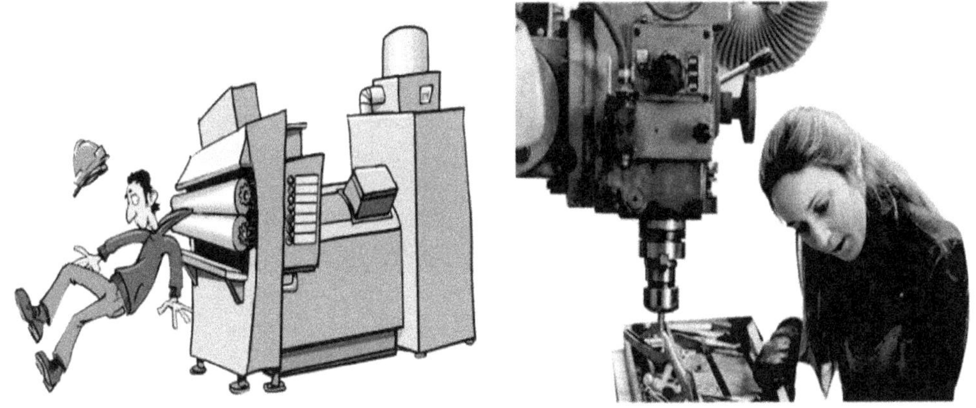

Figure 1.4 Dangers posed by loose clothing and loose hair

1.3.5 Warning symbols

Warning symbols convey a message through pictures. The warning symbols have been classified based on their color: red, yellow, blue, and green. The red color displays prohibition or danger alarm, the yellow color is used to indicate warning, mandatory instructions are displayed using the blue color, whereas the green color indicates important information, such as first aid and emergency exit. A worker in a workshop bears the risk of an injury due to carelessness and/or lack of identification of warning symbols. Hence, it becomes important to know the warning symbols and their meaning before starting work in the workshop. Figure 1.5 shows common warning symbols used in the workshops.

Figure 1.5 Warning symbols and their description

1.4 SUMMARY

In this chapter, the discussion was focused on the introduction to workshop practices to provide an overview of their applications. Different types of workshop practices have also been discussed in brief, providing information about the scope of this manual. Another important aspect of this chapter is to introduce safety practices that need to be followed by the learners.

QUESTIONS

Q1.1 Name the workshop practices or the manufacturing processes that may have been used to produce the following products.

    i.   Structural frame of a bicycle

    ii.   Steel frame or bookshelf near to you (in office or home)

    iii.   Wooden stool for sitting

    iv.   Switch board in your room

    v.   The dumbbells used for weightlifting

    vi.   Nuts and bolts available in a product available near you.

    vii.   Table available near you (study/center/dining, etc.)

    viii. Almirah available near you

Q1.2 What are the important safety precautions to be followed for the workshop practice?

Q1.3 What are the dangers and safety hazards if the safety precautions are not followed?

Q1.4 Look around to find illustrative posters and signboards related to safety in the workshops or elsewhere and state briefly about them.

Chapter 2

CARPENTRY SHOP

2.1 INTRODUCTION

Carpentry shop is used for carrying out a number of carpentry operations to produce wooden products, such as windows, doors, building floors, furniture, interior decoration, toys and construction of roofs. In carpentry shop, making joints is one of the most important tasks required to form a finished product. Therefore, the primary objective of this chapter is to apply 'learn by practice' approach for making carpentry joints. The following sections discuss tools, joint types, and safety measures in relation to the carpentry shop.

2.2 TIMBER

Timber from mature trees is the most important raw material for carpentry. The trees are first sawed into different sizes to suit the intended use. The pattern or appearance of wood cell fibers on the cut surfaces is referred to as "grain." The wood must be milled with the grain running parallel to the direction of the cut.

2.2.1 Wood classification

The two main types of wood are hardwood and softwood. The contrasting properties of soft and hardwood are shown in Table 2.1.

Table 2.1 Properties of hardwood and softwood

Softwood	Hardwood
Light in color, weight and less durable	Dark in color, heavy in weight and durable
Easier to be worked on and split	Difficult to work on and split
High tensile strength but weak shear forces.	High resistance to both tensile and shear forces.
Examples of softwood are kair, deodar, conifers, chir, seemal and walnut.	Examples of hardwood are teak, babul, sal, beach, ash mango, oak, neem and shisham.

2.3 TIMBER SIZES

Timber is classified according to the variety of sizes and shapes. Following is the classification of timber.

i. *Log* (Figure 2.1 (a)) is the trunk of a dead tree without any tree branches. Length of a log is generally more than 6 feet (1.8 meters).

ii. *Balk* (Figure 2.1 (b)) is a roughly square-shaped piece of wood with cross-section dimensions bigger than 50 mm x 50 mm, and a length of more than 200 mm. A balk is produced by removing bark and sapwood from a log.

iii. *Post* (Figure 2.1 (c)) is a rectangular-shaped piece of wood whose thickness lies in the range of 50 mm to 100 mm, whereas its width is between 125 mm to 175 mm.

iv. *Plank* (Figure 2.1 (d)) is a sawn timber piece having a thickness of less than 50 mm with a width greater than 50 mm.

v. *Board* (Figure 2.1 (e)) is a piece of wood having a thickness of less than 50 mm and a width of greater than 150 mm.

vi. *Reapers* (Figure 2.1 (f)) are assorted and non-standard-sized pieces of wood that do not have any specific shape and size.

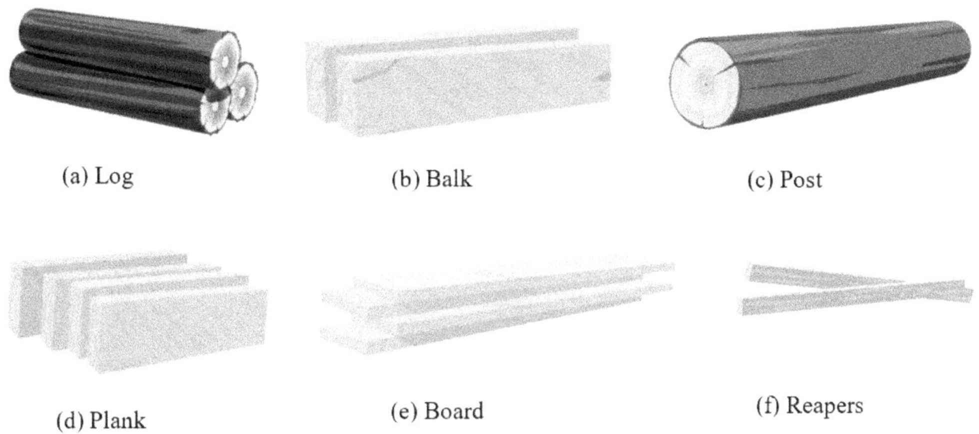

Figure 2.1 Classification of timber

2.4 SEASONING OF WOOD

A freshly felled tree has a high moisture content, which if not controlled, is likely to warp, shrink, crack, or decay the timber made from it. Therefore, seasoning of wood is done to remove excess moisture in the wood which makes the wood durable, lighter and ideal for all carpentry works.

2.5 CARPENTRY TOOLS

Several tools are used in the carpentry shop for different purposes, such as marking, measuring, holding, planning, cutting, drilling, boring and striking. The tools used in the carpentry shop are described in the following sections.

2.5.1 Marking and measuring tools

Marking and measurement tools are essential in the carpentry shop, as they help to transfer dimensions and notations onto the work to make accurate sized parts. The following paragraphs discuss the marking and measuring tools required in a carpentry shop, whereas their illustrations are provided in Figure 2.2.

i. *Steel rule* (Figure. 2.2 (a)) is used for linear measurements. It consists of a long, thin metal strip with a unit division scale marked on it.

ii. *Steel tape* (Figure. 2.2 (b)) is used for inspecting or marking longer dimensions with the help of a flexible measuring tape made of steel.

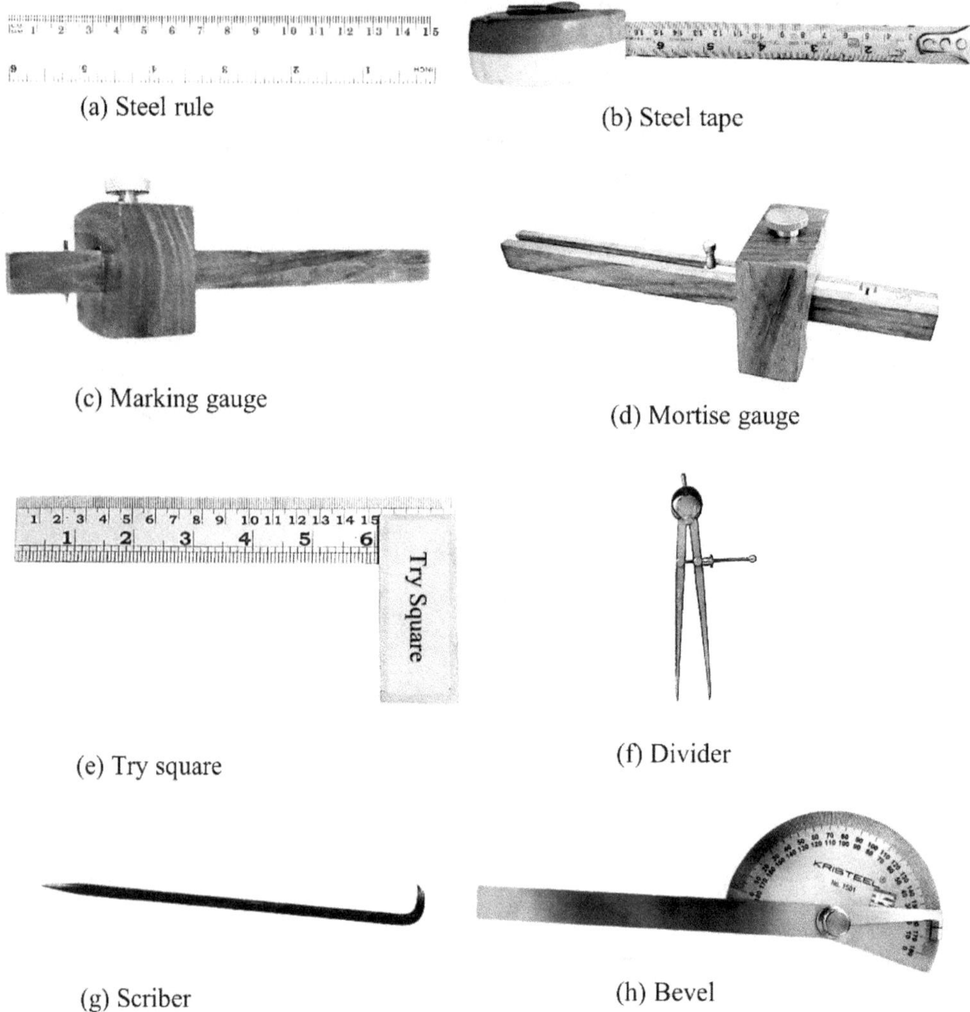

(a) Steel rule

(b) Steel tape

(c) Marking gauge

(d) Mortise gauge

(e) Try square

(f) Divider

(g) Scriber

(h) Bevel

Figure 2.2 Marking and measuring tools used in carpentry

iii. *Marking gauge* (Figure. 2.2 (c)) is used to mark parallel lines to the edge of a wooden workpiece. A sliding wooden stock (head) is mounted on a square wooden stem, whereas a marking pin is attached to the stem. The stock is adjusted to any desired distance from the marking point and fixed in the required position with the help of a screw.

iv. *Mortise gauge* (Figure 2.2 (d)) is also used for marking parallel lines on the stock. It has two sharp edged pins with the distance between them being adjustable.

v. *Try square* (Figure 2.2 (e)) is used for multiple purposes, such as measuring and setting out dimensions, drawing parallel lines at right angles, and testing the squareness and straightness of planed surfaces. It is made up of a cast iron stock and a steel blade fitted in it. The size of a try square ranges from 150 mm to 300 mm, which represents the length of its blade.

vi. *Compass and divider* (Figure 2.2 (f)) are used for marking arcs, circles, divisions, and distances on the surface of the wood.

vii. *Scriber* (Figure 2.2 (g)) is used for marking on the timber. A scriber has two ends, one of which is pointed, and the other has a sharp cutting edge.

viii. *Bevel* (Figure 2.2 (h)) is used for checking angles and bevels. The body of a bevel consists of an adjustable blade, which can be changed between 0 to 180 degrees to the stock.

2.5.2 Holding tools

A variety of holding equipment are used to hold workpieces in the carpentry shop, which are discussed in the following paragraphs.

i. *Carpenter's vice* (Figure 2.3 (a)) is a commonly used work holding device in a carpentry shop. It's one jaw is fixed to the table while the other can be moved with the help of a screw and a handle.

ii. *C-clamp* (Figure 2.3 (b)) is used for holding small workpieces.

iii. *Bar cramp* (Figure 2.3 (c)) is made of a steel bar of T-section, with malleable iron fitting and a steel screw.

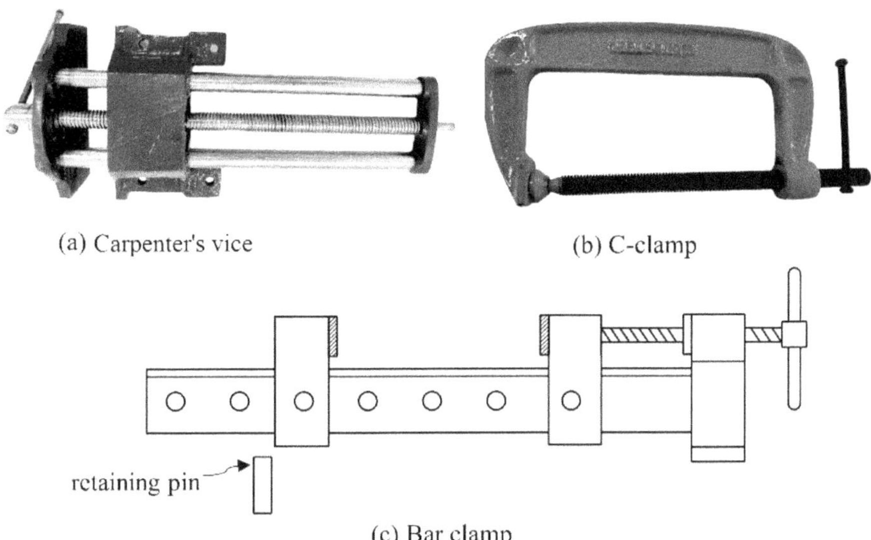

(a) Carpenter's vice (b) C-clamp

(c) Bar clamp

Figure 2.3 Holding tools

2.5.3 Planning tools

Wood surfaces can be made flat with the help of planning tools. The cutting blade of a planner is very similar to a chisel, which is fitted at an angle into a block of metal or wood. The following paragraphs discuss planning tools for a carpentry shop.

i.   *Jack plane* is the most common general purpose planning tool used in a carpentry shop. It includes a wooden body, a handle to hold the planer during operation, a high carbon steel cutting blade, a cap iron to provide reinforcement against the cutting forces, and a wedge. Its length is about 35 cm. The cutting blade should be inclined at 45 degrees and its cutting edge has a slight curvature. It is used for oblique planning and quick removal of material on rough work. Figure 2.4 (a) shows snapshots of a wooden jack plane.

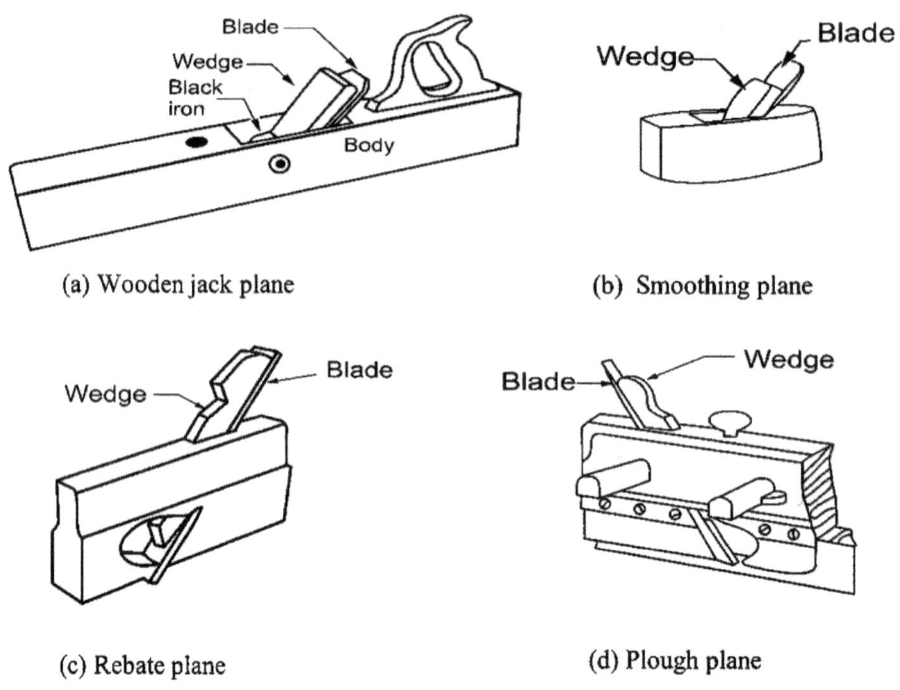

Figure 2.4 Planning tools

ii.   *Smoothing plane* (Figure 2.4 (b)) is a small wooden jack plane used for finishing work. The blade of a smoothing plane has a straight cutting edge, and it is about 20 cm to 25 cm long. Being short in size, it can follow even the slight depressions in the stock. Hence, smoothing plane is used for getting better finish than the jackplane. Therefore, smoothing plane is normally used after using the jack plane.

iii.   *Rebate plane* (Figure 2.4 (c)) is a hand plane used for making a rebate in a wooden workpiece. A rebate is a recess along the edge of a piece of wood, which is generally used for positioning glass in frames and doors.

iv.   *Plough plane* (Figure 2.4 (d)) is used to cut narrow grooves, which are used to fix panels in a door.

2.5.4 Cutting tools

Different cutting tools, such as saws, chisels, drills, and boring tools are used in the carpentry shop. The following paragraphs discuss saws, which are very important cutting tools used in the carpentry shop. Rest of the cutting tools are discussed in the subsequent sections. A saw is used to cut wood into pieces. There are different types of saws, designed to suit a variety of purposes. A saw is specified by the length of its toothed edge. Figure 2.5 shows the most commonly used saws in the carpentry shop.

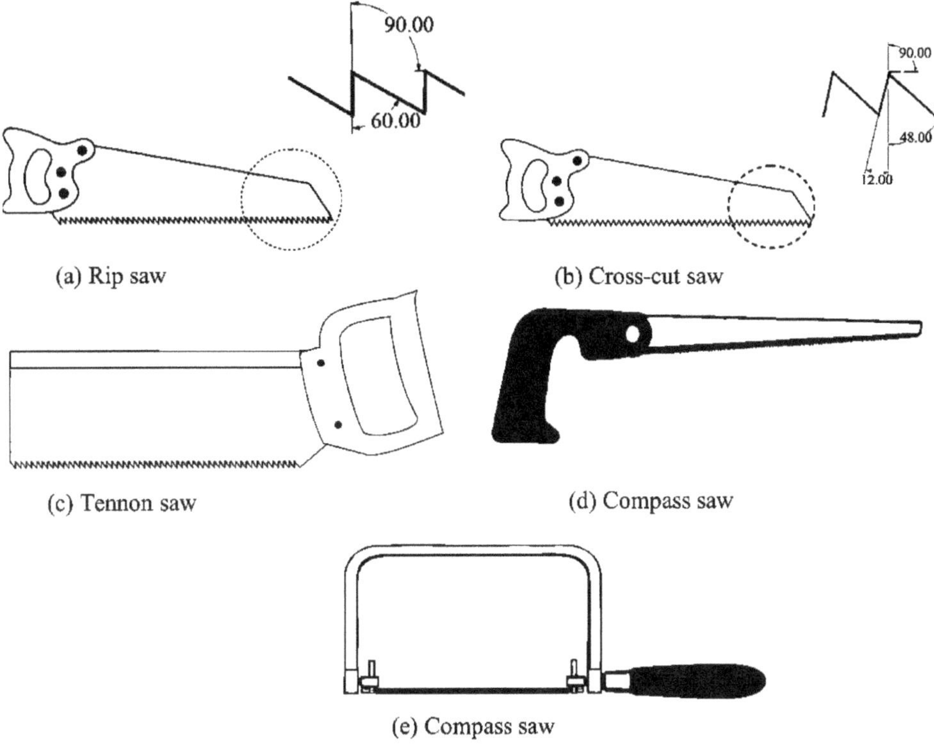

Figure 2.5 Carpentry saws

i. *Rip saw* is used for cutting the stock along the grains. The cutting edge of this saw makes a steeper angle, i.e., about 60° (Figure 2.5 (a)).

ii. *Cross-cut saw* is used to cut across the grains of the stock. Its teeth are so set that the thickness of the cut made will be wider than the blade thickness, which allows the blade to move freely in the cut without sticking. Crosscut saw makes an angle of about 48° (Figure 2.5 (b)) with the surface of the stock.

iii. *Tenon saw* (Figure 2.5 (c)) is used for cutting stock, either along or across the grain. It is employed in fine cabinet work and for cutting tenons for tiny and delicate cuts. Since the blade of this saw is very thin, a thick back steel strip is used to stiffen it. Therefore, it is sometimes referred to as a backsaw. Tenon saw teeth resemble cross-cut saw teeth in shape.

iv. *Compass saw* (Figure 2.5 (d)) is used for heavy work and has a narrow, longer, but stronger tapering blade. This saw's blade is equipped with an open-style wooden handle.

v. *Coping saw* (Figure 2.5 (e)) makes use of a small blade, which is tightly secured from both of its ends with the help of a rigid frame. This is generally used for cutting curves in wood.

2.5.5 Chisels

Chisels are used to precisely cut and shape wood (Figure 2.6 (a)) with the help of a sharpened edge also called the blade. The blade width for carpentry chisels may range from 3 mm to 50 mm. Furthermore, in addition to the blade widths, chisels come in a variety of blade lengths. The chisel blades are generally made of tool steel or forged steel. Most of the carpentry chisels are tang-style, with a steel shank that slides inside the wooden handle. The following paragraphs discuss various chisel types.

i. *Firmer chisel* is one of the strongest chisels (Figure 2.6 (b)), which is so versatile that it can be used either with a mallet or by hand to apply pressure. A firmer chisel has a flat blade that comes in a variety of widths, from 3 mm to 50 mm.

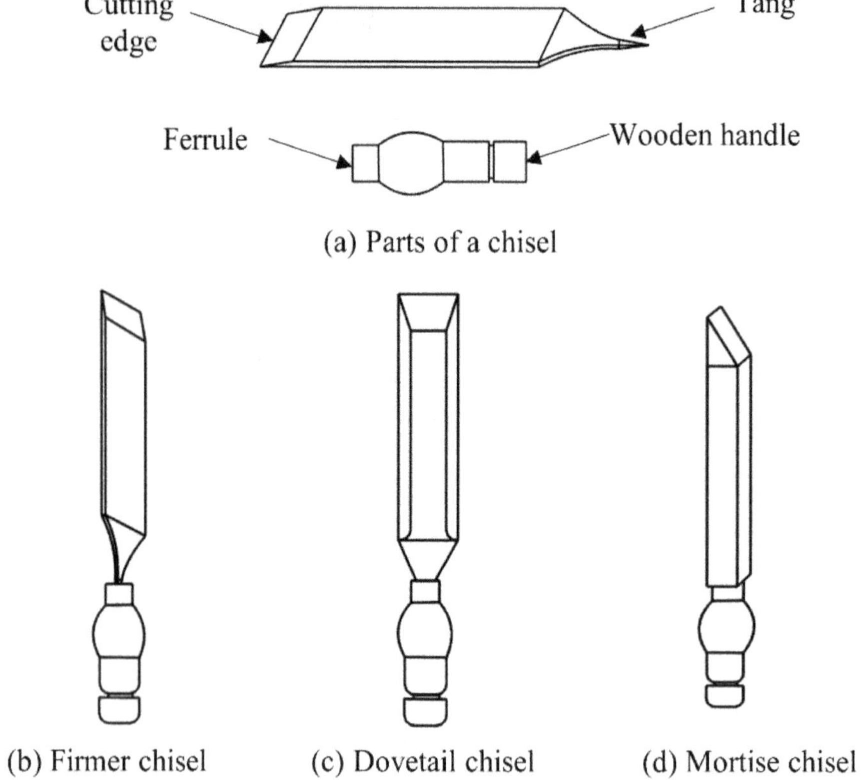

Figure 2.6 Chisel parts and types of chisels

ii. *Dovetail chisel* with a back-beveled blade is depicted in Figure 2.6 (c). It has sharp comer entry points for finishing.

iii. *Mortise chisel* (Figure 2.6 (d)) is used for creating mortises and chipping the interior of holes. Cross-section of the mortise chisel is sized to withstand powerful blows during mortising. Furthermore, the cross-section of a mortise chisel is even stronger closer to the shank.

2.5.6 Drilling and boring tools

Making round holes in wood requires drilling and boring tools, which are selected based on the nature and function of the hole. The most popular drilling and boring tools are discussed in this section and their schematics are shown in Figure 2.7 (a)- (d).

i. *Carpenter's brace* (Figure 2.7 (a)) is used for rotating auger bits and twist drills to produce holes in the wood. Using this tool, holes may be made in a corner where a complete revolution of the handle cannot be made. The size of a brace is determined by its sweep.

ii. *Auger bits and drills* (Figure 2.7(b)) are the cutting tools that are clamped to either the carpenter's brace or hand drill. Its cutting edges resemble a twist drill which are used for drilling large diameter holes with the help of hand pressure.

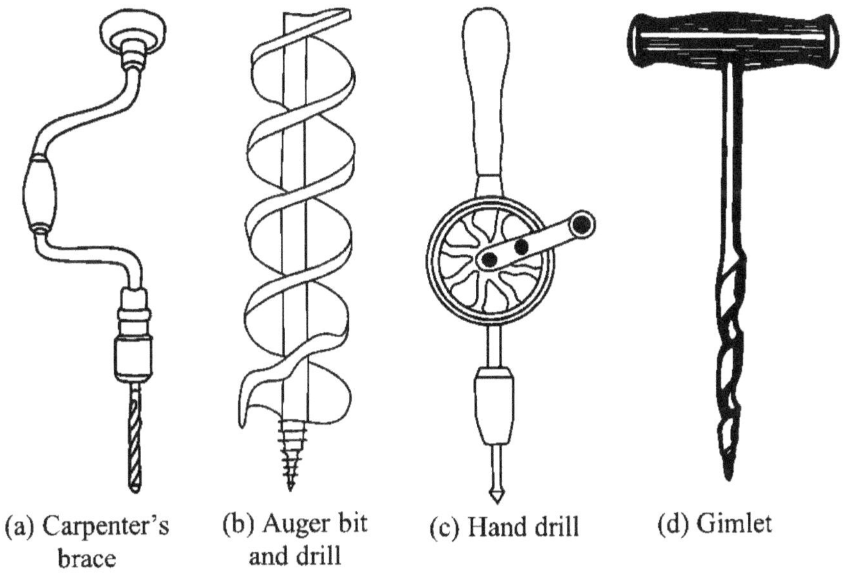

(a) Carpenter's brace   (b) Auger bit and drill   (c) Hand drill   (d) Gimlet

Figure 2.7 Drilling and boring tools

iii. *Hand drill* (Figure 2.7 (c)) is used for drilling small holes with the help of a straight shank drill. It is small, light in weight and more convenient to use than a brace. The drill bit is clamped in the chuck at its end and is rotated by a handle attached to the gear and pinion arrangement. During drilling, the drill bit is guided into the wood, necessitating only moderate pressure on the brace. The helical flutes on its surface carry the chips to the outer surface.

iv. *Gimlet* (Figure 2.7 (d)) is also used for drilling large diameter holes with the application of rotations and hand pressure.

2.5.7 Striking tools

Striking or impelling tools are used for driving chisels and nails into the wood and for assembly work. The two types of most commonly used striking tools used for woodworking are mentioned below.

i. *Mallet* is a small hammer of a round or rectangular cross-section, which is generally made of hardwood. A mallet is used to give light blows to cutting tools with wooden heads such as chisels and gouges. Figure 2.8 (a) shows a snapshot of a mallet.

ii. *Hammer* used in carpentry is of two types, cross peen hammer and claw hammer. The cross-peen hammer (Figure 2.8 (b)) is used for light bench work. It has a flat cast steel head on one end and a sharp peen other end. The handle is made of either wood or bamboo. The claw hammer (Figure 2.8 (c)), in addition to being used as a hammer can also be used for pulling out bent nails and for this reason is preferred by the woodworkers.

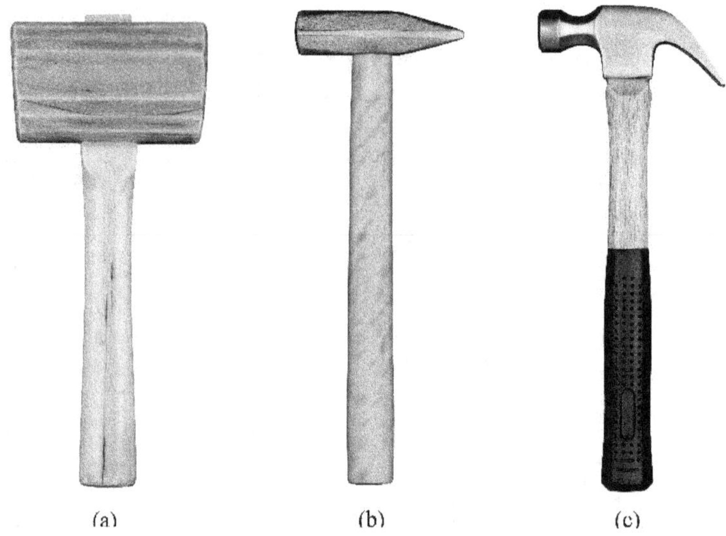

Figure 2.8 (a) Mallet hammer, (b) Cross peen hammer, and (c) Claw hammer

2.5.8 Miscellaneous tools

A carpenter uses a few miscellaneous tools such as rasps, pincer, and screwdrivers in the carpentry shop. These tools are described below.

i. *Rasps and files* (Figure 2.9 (a)) are used to clean and shape wood with curved surfaces.

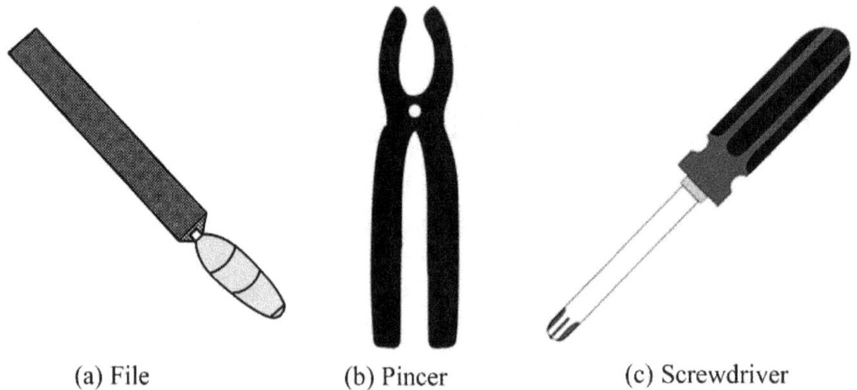

Figure 2.9 Illustration of a file, pincer and screwdriver

ii. *Pincer* (Figure 2.9 (b)) is used to pull nails from the wood.

iii.   *Screwdriver* (Figure 2.9 (c)) is used for screwing or unscrewing nuts and screws used in the wood.

2.6 WOOD JOINTS

There are many types of joints used to connect wood stock. Each joint has a definite use and requires planing, cutting and sizing. The strength of a joint depends upon the amount of contact between the mating workpieces. If a joint does not have much contact area, then it must be reinforced with the nails, screws, or dowels. A brief discussion on the commonly used wood joints, such as lap joint and miter joint is given in the following paragraphs.

In *Lap joints*, an equal amount of wood is removed from the two workpieces (Figure 2.10 (a)-(c)). Lap joints are easy to lay out using a try square and a marking gauge. This type of joint is used for small boxes to large pieces of furniture.

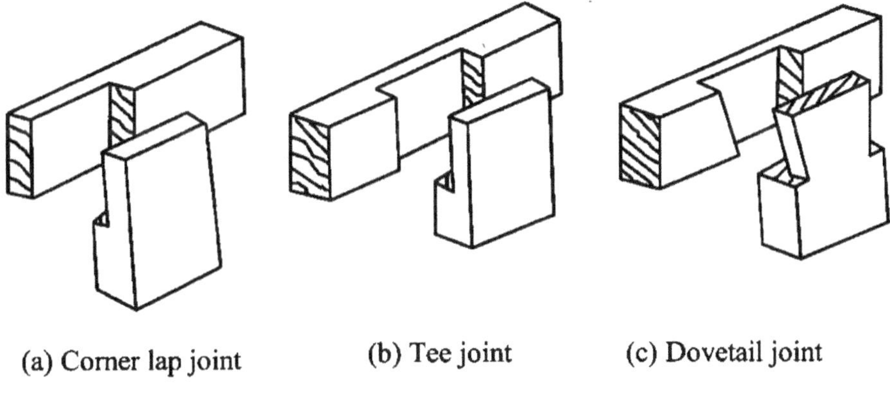

(a) Corner lap joint   **(b) Tee joint**   **(c) Dovetail joint**

Figure 2.10 Lap joints

A mortise and tenon joint is used in the construction of quality furniture. It results in a strong joint and requires considerable skill to make it. Figure 2.11(a)-(c) show different types of mortise and tenon joints.

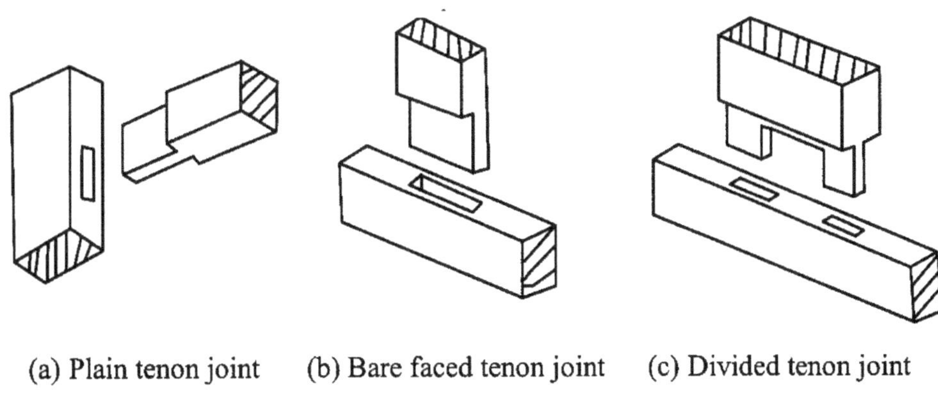

(a) Plain tenon joint   **(b) Bare faced tenon joint**   **(c) Divided tenon joint**

Figure 2.11 Mortise and tenon joints

Bridle joint is just the reverse of the mortise and tenon joint. This joint is used where the members are of square or near square section and unsuitable for mortise and tenon joint. Figure 2.12 shows the types of bridle joints.

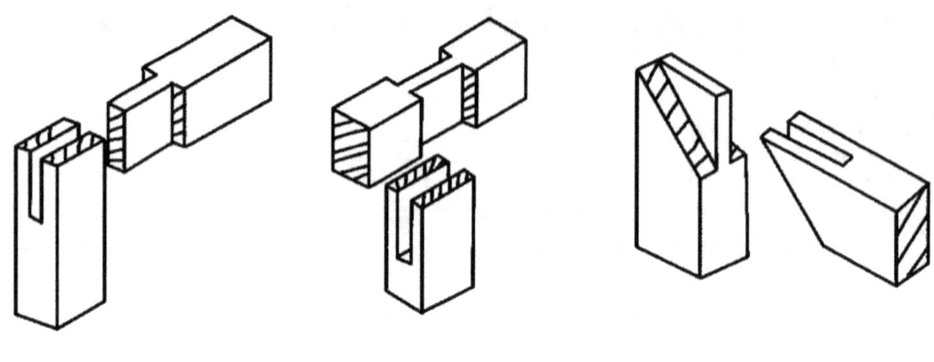

(a) Corner bridle joint   (b) Tee bridle joint   (c) Miter faces bridle joint

Figure 2.12 Bridle joints

2.7 SAFETY PRECAUTIONS FOR CARPENTRY SHOP

In addition to the general safety precautions applicable to workshops, already discussed in Chapter 1, following safety precautions are useful for working in a carpentry shop.

i.   Ensure your hands are not in front of the sharp-edged tools while working with them.
ii. A dull or blunt tool requires excessive pressure (or consumes more power) and causes slip. Therefore, sharpen the tools whenever required.
iii. Take care when you use your thumb as a guide in cross-cutting and ripping.
iv. Don't use wood having nails for any of the carpentry operations as it may damage the tool or may harm the operator.

PRACTICE NO. 2.1

Job: Make a wooden piece of required size using marking, planning, and sawing operations.

Objectives: Understand and learn by practice various carpentry shop operations, like marking, sawing, and finishing, besides respective tool handling.

Machines, equipment, and tools required: Steel foot rule, try square, iron jack plane, marking gauge and crosscut saw.

Material required: Softwood, such as Kali wood and Red Mirindi wood of size 155 mm x 45 mm x 35 mm, approximately.

Schematic of the job:

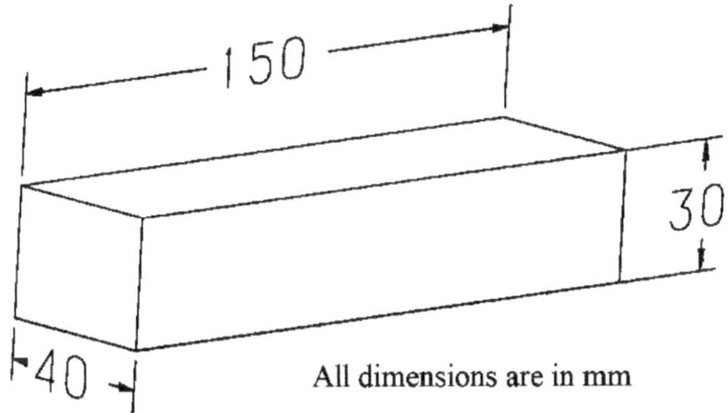

Figure P2.1 Wooden piece

Procedure:

i.   Take a wooden workpiece with approximate size of 155 mm x 45 mm x 35 mm with the help of a crosscut saw.

ii. Hold the wooden workpiece in a clamping vice.

iii. Set the marking gauge and mark all the dimensions in horizontal and vertical directions as per the description given in Figure P2.1.

iv. Use a hacksaw and remove extra material from the workpiece up to the given dimensions in both the horizontal and vertical directions leaving some for planning.

v.   Plane all the required faces using an iron jack plane to size them as per the dimensions given in Figure P2.1.

vi. Fill in the workshop practice response sheet given at the end of this chapter and answer the questions provided therein. Get your instructor's feedback from the workpiece prepared by you.

PRACTICE NO. 2.2

Job: Prepare a wooden Tee Lap joint

Objectives: Learn by practice carpentry shop operations for making a wooden Tee Lap joint.

Machines, equipment, and tools required: Steel foot rule, try square, iron jack plane, marking gauge, crosscut saw, firmer chisel, mortise chisel rasp file, plastic body hammer.

Material required: Softwood (Kali or Red Mirindi) of size 300 mm x 45 mm x 35 mm

Schematic of the job:

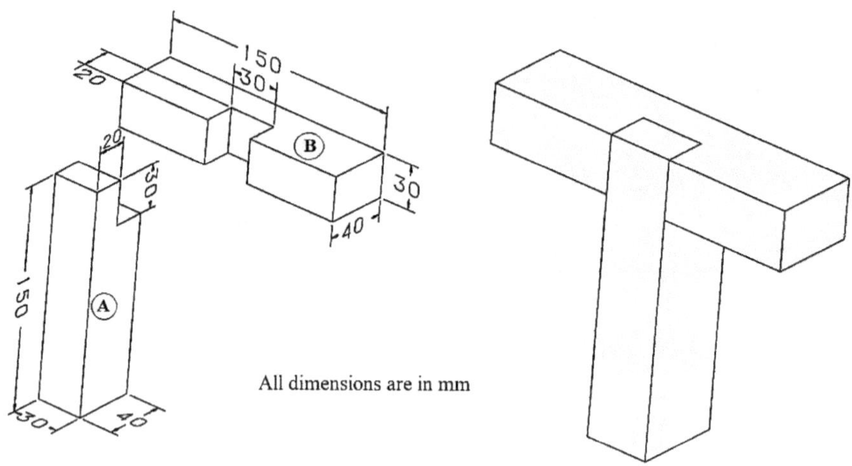

Figure P2.2 Tee joint

Procedure:

    i.   Select a wooden workpiece of about 300 mm x 45 mm x 35 mm in size. Measure all the dimensions with a steel foot rule.

    ii.   Prepare a reference surface with the help of an iron jack plane by removing the material from one long surface. Remove as little material as much possible.

    iii.   The job faces are checked for squareness using a try square.

    iv.   Use a marking gauge to mark the thickness and width of the wooden blocks to be 40 mm and 30 mm, respectively.

    v.   With the help of a crosscut saw, the wooden workpiece is cut into two equal workpieces, each having a length of 150 mm, designated as 'A' and 'B', respectively.

    vi.   The mating dimensions of workpieces 'A' and 'B' are then marked using a ruler and a marking gauge.

    vii.   With a crosscut saw and a chisel, mating portions of both the workpieces 'A' and 'B' are marked with the help of a steel foot rule and a marking gauge. Referring to Figure P2.2, you may notice that the mating dimensions have a size of 30 mm x 30 mm x 20 mm.

    viii.   The mating dimensions of both the workpieces are chiseled and sawed to size. It should be taken care that the parts are fitted to obtain a slightly tight joint.

    ix.   Finishing is given to the parts if required so that proper fitting is obtained.

    x.   Fill in the workshop practice response sheet given at the end of this chapter and answer the questions provided therein. Get your instructor's feedback on the workpiece prepared by you.

PRACTICE NO. 2.3

Job: Prepare a wooden Tee Bridle joint

Objectives: Learn by practice, carpentry shop operations and tools for making a Tee Bridle joint.

Machines, equipment, and tools required: Engineering steel foot rule, try-square, marking gauge, iron jack planer, firmer chisel, mortise chisel, rasp file.

Material required: Soft wood (Kali or Red mirindi) of size (300 mm x 50 mm x 50 mm) approximately or two wooden pieces of 150 mm x 50 mm x 50 mm.

Schematic of the job:

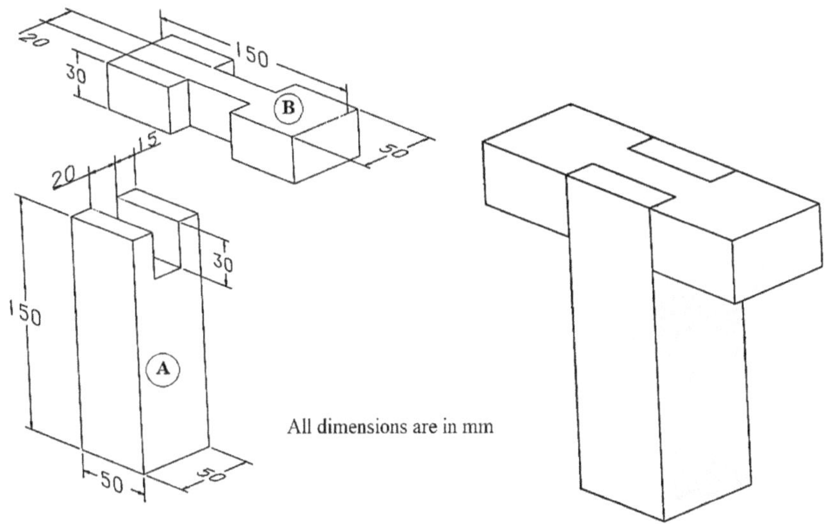

Figure P2.3 Tee bridle joint

Procedure:

    i.   Cut a wooden piece of size 300 mm x 50 mm x 30 mm with the help of crosscut saw.

    ii.   Prepare a reference surface with the help of an iron jack plane by removing the material from one long surface of the workpiece. Remove as little material as much possible.

    iii.   Remove extra material with the help of an iron jack plane and cut the workpieces into two pieces of size 150 mm x 50 mm x 30 mm each, designated as 'A' and 'B'.

    iv.   Check flatness and perpendicularity of the surfaces with the help of a try square.

    v.   On workpiece 'A', do the marking for cutting 30 mm x 15 mm x 50 mm (see Figure P2.3 for descriptions) for removing the material with the help of crosscut saw, mortise and firmer chisel, and hammer.

    vi.   Set the marking gauge at 30 mm and mark parallel lines on the opposing sides of the workpiece 'B'.

    vii.   Further, mark the workpiece 'B' for 50 mm x 15 mm x 30 mm size on the two opposing sides as shown in the figure for removing the material. Remove the material with the help of a chisel and a mallet.

    viii.   Assemble the workpiece 'A' and 'B' to get a wooden Tee bridle joint.

    ix.   Fill in the workshop practice response sheet given at the end of this chapter and answer the questions provided therein. Get your instructor's feedback on the workpiece prepared by you.

PRACTICE NO. 2.4

Job: Make a wooden mortise and tenon joint

Objectives: To learn by practice carpentry shop operations, materials, tools, joints, and safety precautions for preparing a mortise and tenon joint.

Machines, equipment, and tools required: Steel foot rule, marking gauge, try square, wooden jack plane, hacksaw, rasp file, firmer chisel, mortise chisel and C-clamp.

Material required: Soft wood (Kali or Red mirindi) of size (300 mm x 45 mm x 50 mm)

Schematic of the job:

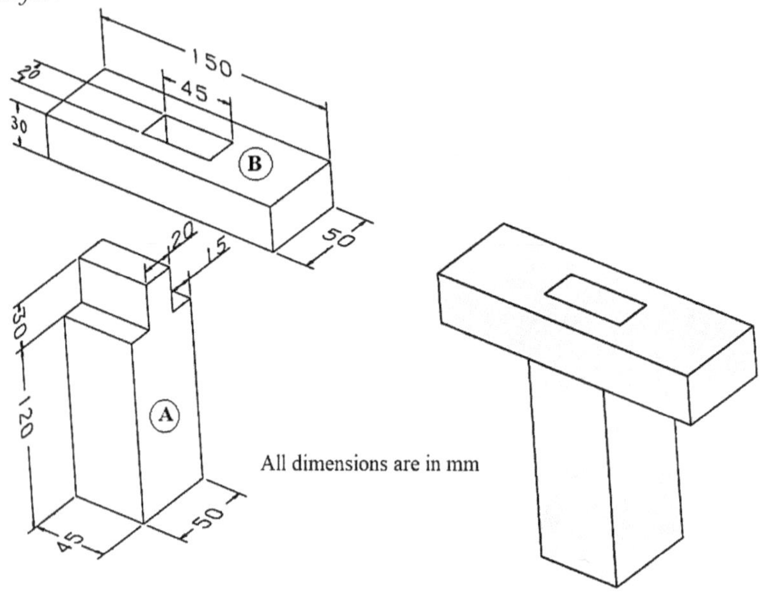

Figure P2.4 Mortise and tenon joint

Procedure:

i. Select a wooden workpiece of size approximately 300 mm x 45 mm x 50 mm.

ii. Plane the workpiece with the help of a wooden jackplane to maintain the required dimensions.

iii. Fix the workpieces in a carpentry vice and cut them into two equal pieces, each of size 150 mm in length with the help of a hacksaw. Mark them 'A' and 'B', respectively

iv. Mark the two workpieces for cutting as per the dimensions shown in Figure P2.4.

v. Fix the workpieces one by one by keeping it on the bench with the help of a C-Clamp.

vi. Work on the workpiece 'A' to cut and finish it to size as per the dimensions shown in Figure P2.4 using hacksaw, chisel, and rasp file.

vii. With the help of a chisel and mallet, make a groove in workpiece 'B' as per the dimensions shown in Figure P2.4.

viii. The two workpieces are fitted to make the mortise and tenon joint as per drawing.

ix. Fill in the workshop practice response sheet given at the end of this chapter and answer the questions provided therein. Get your instructor's feedback on the workpiece prepared by you.

WORKSHOP PRACTICE RESPONSE SHEET

Learner's/Student's name: Roll Number:

Job Name: Job material:

Objective of this practice:

Which operations and tools/equipment you used to complete the job?
(To be mentioned as per the sequence)

Operations   *Tools and Equipment used*   *Illustration*

Describe what did you learn in this practice session?

State the safety precautions followed by you.

Carefully inspect the job and note shortcomings/defects (in consultation with the instructor)

Instructor's remarks

Date: Signed/Checked

Chapter 3

FOUNDRY SHOP

3.1 INTRODUCTION

Foundry or casting is the process of making metallic jobs of required shapes by melting the metal in a furnace followed by pouring it into a collapsible mold. The mold is a cavity created in the molding sand by using a pattern of the job to be made. The metal is allowed to solidify, and the mold is later broken to get the solidified part out of it. A pattern is a replica of the part to be made by the casting process.

The procedure for metal casting has the following major steps,

    i.   Prepare the mold by covering the pattern with the molding sand and ramming it to give it a given shape.

   ii.   The pattern is withdrawn to generate a cavity in the molding sand, the sand containing the cavity is also called the mold.

  iii.   Pouring the molten metal into the mold.

  iv.   The solidified metal which takes the same shape as that of the mold is taken out by breaking the mold.

The process of making parts in the foundry shop is also known as the casting process. Following sections further explain the advantages and disadvantages of the foundry shop and casting process.

3.1.1 Advantages of the casting process

The casting process has the following advantages over other manufacturing processes.

    i.   It can create parts with complex shapes very economically.

   ii.   Size of parts can be small to very large.

  iii.   The parts produced generally have high compressive strength.

  iv.   Wide range of metals and alloys can be cast, such as aluminum, zinc, and cast iron.

   v.   It produces parts that possess isotropic structures and properties.

  vi.   It is found to be very cheap, particularly when production volume is low.

 vii.   It is also capable of producing composite components.

3.1.2 Disadvantages of the casting process

The casting process has the following disadvantages.

i.   It gives a poor surface finish and therefore requires additional finishing operations.
ii.   Defects are often found in the parts produced using the casting process.
iii. The parts produced using the casting process generally have lower fatigue strength when compared with those produced using the forging process.
iv.   The casting process is generally not economical for mass production.

3.2 TYPES OF PATTERNS

A discussion about the most used patterns in the foundry shop is made in the following paragraphs along with their illustrations.

3.2.1 Single piece pattern

The single-piece solid pattern behaves as one piece and may be considered as the simplest form of pattern. Figure 3.1 (a) shows a snapshot of a single piece pattern.

3.2.2 Split or two-piece pattern

A solid pattern splits into two pieces to facilitate withdrawal of the pattern from the mold. The two pieces of the split pattern are joined at the parting line using dowel pins. Figure 3.1 (b) shows the snapshot of a split pattern.

3.2.3 Gated pattern

In mass production, many castings are produced in a single multi-cavity mold by connecting multiple patterns. In this kind of multi-cavity mold, the gates or runners for the molten metal are made by connecting parts between the individual patterns. Figure 3.1 (c) shows a snapshot of a gated pattern.

3.2.4 Loose piece pattern

Sometimes, due to the typical shape of the part to be made, it becomes difficult to prepare the mold using a single piece or two-piece pattern. Therefore, for parts of relatively complex shapes, a pattern is made as an assembly of loose pieces. The arrangement of loose pieces of the pattern is so made that their removal from the mold becomes easier. Figure 3.1 (d) shows a snapshot of a loose piece pattern.

3.2.5 Sweep pattern

Sweep patterns are commonly used to prepare large sized symmetrical molds. It has a wooden sweep board, which is attached to a spindle. Outer edge of the sweep board has the same shape as required from the part to be produced. To prepare the mold, the spindle of the sweep board is held vertically, and the sweep board is rotated in the molding sand to create the cavity. Figure 3.1 (e) shows a snapshot of a sweep pattern.

3.2.6 Match plate pattern

Metal match plate patterns are widely used for the mass production of small size castings. Like the split pattern, a match plate pattern is also made of two pieces, however, they are mounted on two sides of a metal plate. (Figure 3.1 (f)).

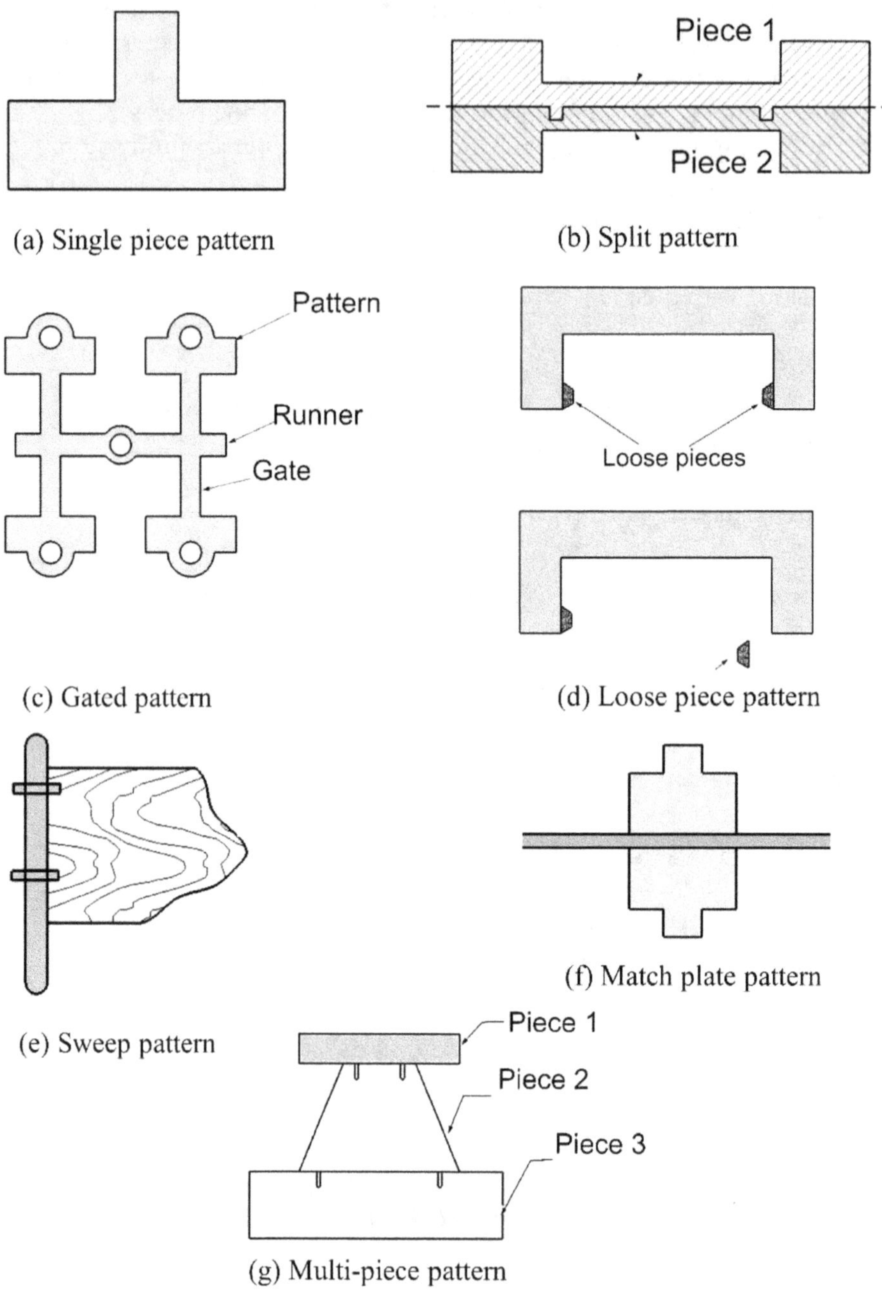

Figure 3.1 Pattern types

3.2.7 Multi-piece pattern

Sometimes, the design of a part is such that it cannot be molded in two parts. Therefore, the pattern needs to be divided into more than two parts to facilitate easy withdrawal of the pattern from the mold. Multi-piece pattern is suitable for such parts that require the pattern to be divided into multiple pieces. A schematic of a multipiece pattern has been shown in Figure 3.1(g). The multi-piece pattern usually requires three molding boxes.

3.3 MOLDING SAND

Molding sand tends to pack well and retain its shape when moistened, compressed, oiled, or heated. It is used to prepare the collapsible mold cavity, which means that after the casting process is complete, the mold can be easily broken without damaging the part.

3.3.1 Composition of molding sand

Molding sand is composed of silica sand, binder, additives, and water. Further details of molding sand are described below.

i. *Silica sand* is the major constituent of the molding sand. It is produced when either quarry stone or granite stone is broken down into small particles. The major advantages of using silica sand in the molding sand are that it is easily available and exhibits the properties of refractoriness, permeability and chemical resistance. Specifications of silica sand include grain size and shape.

ii. *Binder* when added in the silica sand provides sufficient strength and cohesiveness, which helps in maintaining its shape. The most common binders used in molding sand are molasses, linseed oil, clay, sodium silicate and resins.

iii. *Additives* are materials which are mixed into sand to improve its properties. The most commonly used additives are coal dust, wood flour, molasses, corn flour, and pitch.

iv. *Water* is added to the molding sand to improve its bonding action as it penetrates the clay mass present in the molding sand to form microfilm. The maximum thickness of microfilm that clay can hold, determines its bonding capacity. In general, recommended water percentage varies between 2 to 8 percent in the molding sand.

3.3.2 Types of molding sand

Variety of molding sands are available, which are used for different applications. Following paragraphs discuss different kinds of molding sands.

i. *Green molding sand* is composed of silica sand mixed with about 20% of clay and 8% of water. Green sand contains enough moisture to be sufficiently adhesive in its natural state. The major advantage of green molding sand is that it readily takes the shape of the pattern when squeezed and pressed against it.

ii. *Dry sand* is produced by removing the moisture from a green sand mold. Therefore, the mold made using the dry sand is also called a dry sand mold. The dry sand mold thus obtained becomes stronger and compact. Therefore, when the molten metal is poured into the mold, the dry sand is subjected to the most severe conditions as it comes in direct contact with the molten metal. Due to this reason, it is expected that the dry sand must possess good strength and resistance to heat (also called refractoriness). The dry sand is composed of powdered silica sand and clay.

iii. *Loam molding* sand is a mixture of silica sand, clay, and water. The quantity of clay is significantly high to about 50 percent, whereas water content is about 20 percent. The loam sand is applied to give proper shape to the mold by plastering it on the soft bricks, which harden after drying. A major application of the loam sand is for large castings, such as the bed of a lathe machine.

iv. *Backing sand* is re-usable sand obtained from the used mold after the casting process is over. It is also known as the black sand due to its black colour, which is the result of coal dust and burning of sand.

v. *Parting sand* is the fine dry silica sand which is normally used to prevent sticking of the mold halves.

vi. *Core sand* is made by mixing of the silica sand, core oil and other binding materials. Normally, linseed oil, light mineral oil, and resin are used as the core oils. However, in special cases, when large cores are required, pitch (or flour) and water may be added to reduce the core cost.

3.4 TOOLS AND EQUIPMENT

The following tools are the most commonly used in a foundry shop.

i. *Hand riddle* (Figure 3.2 (a)) is made of a wire mesh, which is provided with a circular boundary of the steel sheet. The main application of a hand riddle is to clean the molding sand by removing unwanted material, such as nails or stones. However, hand riddle is used for smaller quantity of sand whereas for larger quantity, power operated riddles may be used.

ii. *Shovel* (Figure 3.2 (b)) is made of a steel pan, which is provided with a wooden handle shovel and is provided with a wooden handle. It is used to mix and condition smaller quantity of moldings sand. It is also used for shifting sand to the molding box.

iii. *Rammer* (Figure 3.2 (c)) is used to strike the molding sand for packing it around the pattern. Rammers help to uniformly pack the molding sand to make it dense and strong. A rammer generally has two ends; the thick round end is known as the butt and the other relatively pointed end is known as the peen.

iv. *Trowels* (Figure 3.2 (d)) are made of flat steel sheet and provided with a handle for finishing of flat surfaces of the mold.

v. *Lifter* (Figure 3.2 (e)) is used for finishing and repairing of the mold, which may get damaged during the mold making process. Lifter is also used to remove loose sand from the mold.

vi. *Sprue pin* (Figure 3.2 (f)) is a tapered piece of wood or other suitable material, which is used for making of a hole in the cope so that the molten metal can be poured into the mold cavity.

vii. *Wood or iron strike off bar* (Figure 3.2 (g)) is used to remove excess sand from the top of the rammed molding box. It has a beveled edge and a perfectly smooth and flat surface.

viii. *Vent wire* (Figure 3.2 (h)) is a piece of wire made of steel or a suitable material, which has a pointed edge at one end. Other end of the vent wire either has a wooden handle or a bent loop of wire for holding. The vent wire is used for making vents after the mold is fully packed with the molding sand by ramming. The vents, which are very narrow holes help escape steam and gases formed during the casting process, through the mold walls.

Figure 3.2 Hand tools used in foundry

ix.   *Slick* (Figure 3.2 (i)) is a small sized tool made of steel which is used for repair and finishing of the mold faces and edges. It has two ends, with different shapes to suit the mold faces needing repair.

x.   *Swab* (Figure 3.2 (j)) is a kind of a brush which has very soft bristles. It is used for providing moisture to the surfaces of the mold which are in contact with the pattern. A secondary application of swab is that it can be used for removing very small sand loose particles from the surface of a mold.

xi.   *Gate cutter* (Figure 3.2 (k)) is made of steel sheet which is used to cut runners and gates in the mold for connecting the sprue hole to the mold cavity.

xii.   *Bellows gun* (Figure 3.2 (l)) is hand operated leather made device that blows (or pumps) air to remove loose sand particles from the mold cavities. Removal of such loose sand particles is required as they may intrude into the molten metal, thus giving rise to defects.

xiii.   *Draw spike* (Figure 3.2 (m)) is made of a thin steel rod, which has sharp pointed tapered and threaded end. The sharp threaded end is inserted inside the pattern surface for its removal from the mold by applying gentle pull without damaging the mold.

xiv.   *Mallet* (Figure 3.2 (n)) is generally used to loosen the pattern, allowing its easy removal. It is used to strike on with a draw spike.

xv.   *Ladle* is a bucket shaped container, which is lined with the refractory material on its inner surface. It has two handles, one each on left and right for the foundry persons to carry molten metal from the furnace to the mold.

xvi.   *Furnace* is used for melting of the solid pieces of metal and alloy. There are several furnaces available in the market. Pit furnace is the most basic furnace used in the foundry shop, which makes use burning of coal to melt the metal.

3.5 MOLDING BOX

Sand molds are prepared in specifically designed boxes known as flasks or molding boxes. The function of molding boxes is to provide the required rigidity and strength to the molding sand. The boxes are typically made in two parts, which are aligned together with the help of dowel pins. As shown in Figure (3.3), the upper part of the mold is known as the cope and the lower part drag. Depending on the size and purpose of the flask, these flasks can be made of wood or metal.

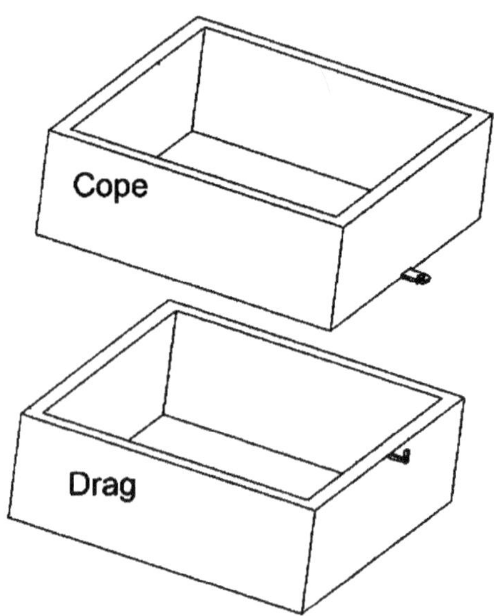

Figure 3.3 A schematic of the molding box

Pattern is used to create the cavity inside the molding box by packing sand around it. The pattern itself is divided along the parting line, which refers to the common surface between the drag and the cope. Sometimes, due to part shape to be molded requires three molding boxes instead of two, known as cope, drag and cheek.

3.6 ELEMENTS OF A GATING SYSTEM

Gating system of a mold facilitates the flow of molten metal from the ladle (which contains the molten metal) to the mold. The main elements needed for the gating system are shown in Figure 3.4 and briefly explained below.

i. *Pouring basin* holds molten metal and acts as a reservoir.

ii. *Sprue*, through which the molten metal flows after pouring is a passage connecting the pouring basin and the runner.

iii. *Runner* allows the molten metal to enter the cavity. It is a horizontal channel that connects the sprue with the gates. Generally, the cross section of a runner is trapezoidal.

iv. *Gate* is the passage that connects the runner with the mold cavity for the flow of the molten metal.

v. *Cope* is the upper part of the molding box.

vi. *Drag* is the bottom part of the molding box.

vii. *Riser* is also known as a feed. It acts as a reservoir to compensate the shrinkage of the molten metal during liquid and solidification stages.

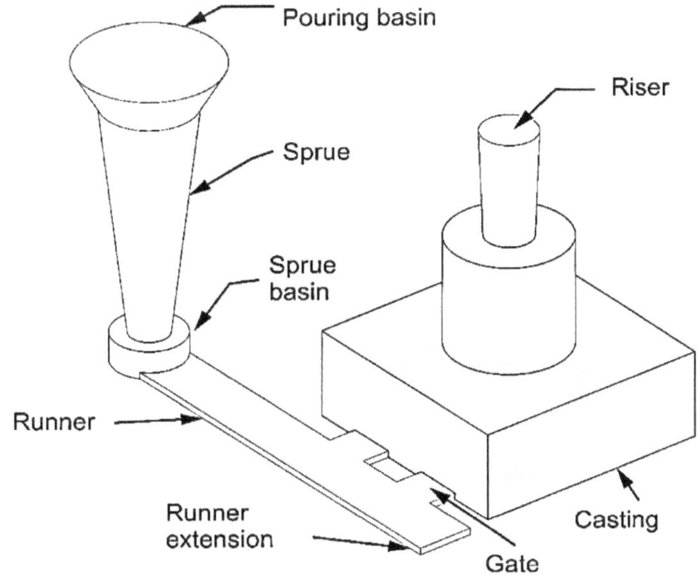

Figure 3.4 Elements of a gating system

3.7 CASTING

The casting process can be divided into the following steps.

i. *Mold preparation*: This step is about packing of sand densely into the mold by placing a pattern of the part in it. Thereafter, a gating system is created by removing the sand followed by removal of the pattern from the sand to create a cavity in the mold. The top and bottom components of a molding box are known as the cope and drag part of the mold, respectively.

ii. *Metal preparation*: In this step a furnace is used to melt the metal. Foundry shop employs different types of furnaces for melting ferrous and non-ferrous materials.

A furnace has a high temperature zone surrounded by a refractory wall structure that can withstand high temperatures while insulating the surrounding area to reduce heat loss. The most common furnaces used in the foundry are pit furnace, reverberatory furnace, induction furnace and crucible furnace. Flux is also used to clean the molten metal of impurities to form slag, which floats on the upper surface of the molten metal. The slag also acts as a heat insulating cover for the molten metal besides preventing its oxidation.

iii. *Pouring and solidification:* In this step the molten metal is poured into the mold through a gating system that runs from the casting cavity through the pouring basin to reach the mold cavity through gating system of the mold. Once the molten metal enters the mold cavity, it cools and solidifies. After the cavity is filled and solidified, the casting takes its final shape, and the casting process is completed.

iv. *Retrieving the part, trimming, and cleaning:* After the solidified metal part cools, it is retrieved by breaking the sand mold. The mold may be dismantled with the help of a vibrating machine that breaks the sand mold and the solidified casting is taken out of the molding box. However, the metal available in the gating system also solidifies and attaches itself to the part. The excess material, which thus gets attached to the solidified part should be trimmed either manually or by using a trimming press. Lastly, the cast part is cleaned from dust and other foreign deposits.

3.8 CASTING DEFECTS

In foundry and casting shop, if proper control is not exercised on the process, it results into unwanted irregularities in the cast part. There are a number of reasons that are responsible for such defects. The common defects are discussed in the following paragraphs and their illustrations provided in Figure 3.5 (a)-(i):

i. *Blow holes* are large and small size round cavities caused due to the presence of gases which displace the molten metal at the cope surface of a casting.

ii. *Sand fusion* is a defect caused by the fusion of sand grains with the molten metal due to which it becomes an integral part of the cast part. The sand diffusion appears as a glassy appearance crust attached to the casting which is brittle too.

iii. *Mismatch* is the misalignment of the cast part due to poor matching of the cope and drag part of the mold cavity.

iv. *Cold lap or cold shut* is caused by the low temperature of the molten metal and its lower fluidity.

v. *Flash or fins* are caused when two halves of the mold are not tightly closed by clamping which results in metal going into the gap between the cope and the drag.

vi. *Crack or hot tears* are caused due to complicated geometry of the part and the mold, which hinders the cast part's contraction, thereby inducing stresses and thin cracks.

vii. *Shrinkage* happens due to inadequate feeding of the molten metal to the cavity. However, risers are used to compensate for such shrinkage of the molten metal during solidification. A riser forms a reservoir of the molten metal and supplies it during solidification of the molten metal to compensate for the shrinkage.

viii.   *Inclusions* are non-metallic impurities which get embedded in the part due to their presence in the molten metal.

ix.   *Porosity* and pin holes are caused due to the reason that during solidification, the gases dissolved in the molten metal try to escape from the melt, forming porosity and pin holes in the cast part.

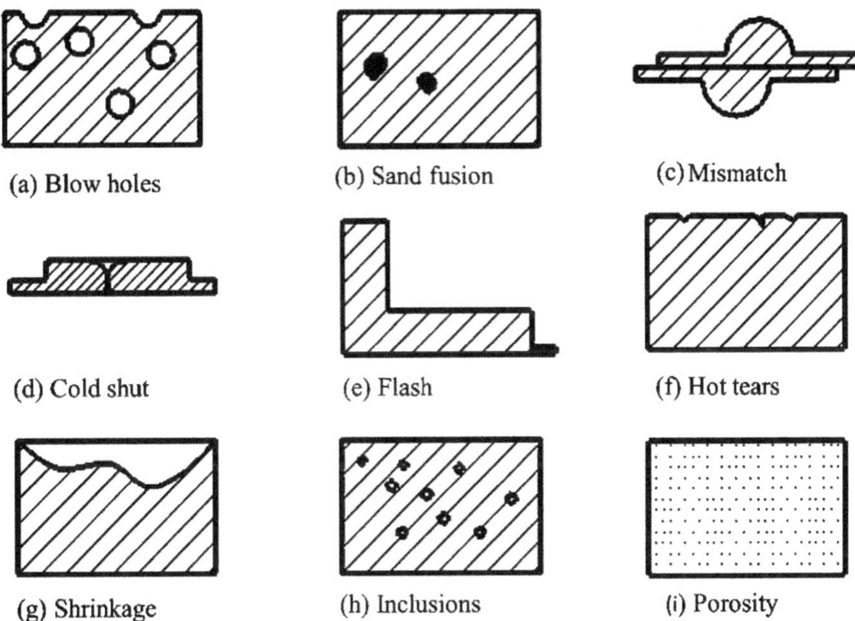

Figure 3.5 Casting defects

3.9 SAFETY PRECAUTIONS

i.   Don't use excess of water in the molding sand due to the reason that water is considered an enemy of the molten metal.

ii.   Provide adequate vents for ventilation on the mold to remove gases and moisture from the mold.

iii.   Never stand near or look over the mold during the pouring of molten metal because it might be too hot.

iv.   The solidified casting takes time to cool. Therefore, the mold or the casting should not be taken in hand for working. This is because any contact with the hot casting may result in second and third-degree burns.

v.   Proper safety gear should be worn during the melting and pouring processes. Examples of such safety gear are heat-resistant gloves, apron shoes and face shield.

vi.   Molten metal spilled on concrete may cause the concrete to explode. Therefore, use a thick sand bed over concrete in case there is a chance of exposure of the molten metal coming in contact with the concrete.

vii.   Always operate in a well-ventilated area. Fumes and dust from combustion, chemical ingredients used, processes, and metals can be toxic.

viii.   Never use a crucible that has been damaged or dropped.

PRACTICE NO. 3.1

Job: Prepare a mold cavity and make the part by the casting process using a solid or single piece pattern

Objectives: Learn by practice the required equipment, tools and operations for preparing a mold using a single piece pattern and making a part using casting process.

Machines, equipment, and tools required: Pit furnace, ladle, molding boxes, single piece pattern, sand rammers, trowel, leveller, vent road, molding sand mixture and other foundry tools.

Material required: Molding sand, clay, water, backing sand, parting sand, aluminum billet and scrap, flux.

Schematic of the job:

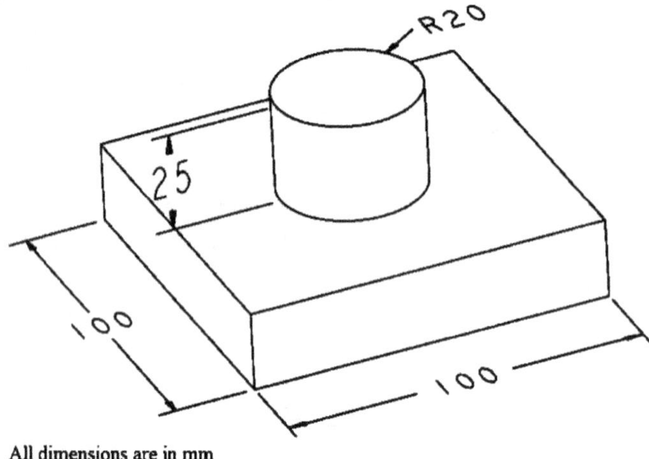

All dimensions are in mm

Figure P3.1 Single piece pattern

Procedure:

i. Prepare the green sand by mixing water and clay with sand.

ii. Place the pattern on the bottom board inside drag half of the molding box. Keep flat side of the pattern on the board (Figure P3.2 (a)).

iii. Sprinkle the parting sand over the pattern.

iv. Fill the drag half of the molding box with the loose green sand around the pattern.

v. Ram the sand packed mold with a rammer by first using the peen end followed by the butt end.

vi. Excess of sand appearing on the top surface of the drag half should be removed using a strike off bar.

vii. The drag half is turned upside down as shown in Figure P3.2 (b).

viii. Loose sand particles are blown away with the help of bellows and upper surface is smoothened.

ix. Cope half of the molding box is placed on the top of the drag half and aligned with it.

x. The riser pin is placed in such a way that it is located on the highest point of the pattern. Also place the sprue pin at about 5 cm to 6 cm away from the pattern. (Figure P3.2 (c)).

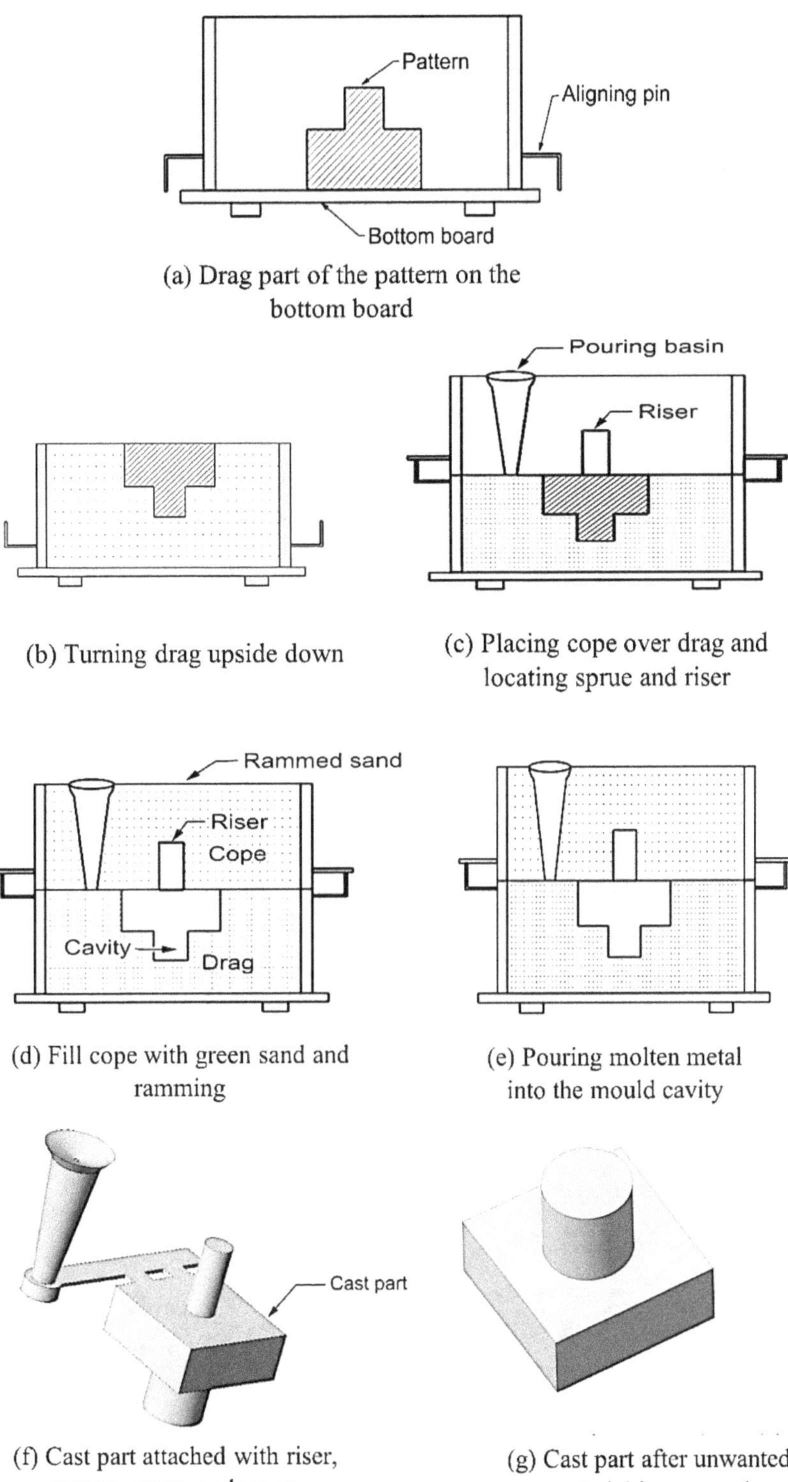

Figure P3.2 Mold preparation and casting using single piece pattern

xi.   The upper surface of the drag half should be visible to you, sprinkle the parting sand on it.

xii.   Completely fill the cope half of the molding box with the green molding sand keeping the position of sprue and riser as such (Figure P3.2 (d)).

xiii.   The excessive sand is removed from top surface of the cope.

xiv.   Pierce vents, which are thin holes through the surface of the cope half with the help of a vent wire.

xv.   Carefully remove the sprue pin and the riser pin from the cope half.

xvi.   Lift the cope and place it aside taking care that it is not damaged.

xvii.   Remove the pattern from the mold by gently rapping it to prevent damage to the mold walls.

xviii.   A funnel shaped hole is made at the top of the sprue hole. This is also called the pouring cup or pouring basin as it is used for pouring of the molten metal. Repair and adjust the mold by adding small amount of the molding sand, if required.

xix.   Prepare a runner and a gate in the drag for making a passage from the sprue base to the mold cavity. Loose particles, if any, are blown off.

xx.   Keep the two mold halves in a secure dry place to allow their drying until they become sufficiently hard to withstand pressure of the molten metal.

xxi.   Close the mold by replacing the cope and placing weights on it.

xxii.   Pour the molten metal into the mold cavity (Figure P3.2 (e)) and allow it to solidify and cool to the room temperature.

xxiii.   Break the mold to take out the casting (Figure P3.2 (f)), clean the casting and remove unwanted material attached to it to get the cast part (see Figure P3.2 (g)).

xxiv.   The casting is visually inspected, and defects are identified by maintaining a record of the same.

xxv.   Fill in the workshop practice response sheet given at the end of chapter 2 and answer the questions provided therein. Get feedback of the instructor on the workpiece prepared by you.

PRACTICE NO. 3.2

Job: Prepare a mold and perform casting operation using a split piece pattern.

Objectives: Learn by practice the process of mold making using a split pattern, metal preparation and melting, pouring the metal into it, reclaim the solid part and identify the defects.

Machines, equipment and tools required: Molding flasks, split (two-piece) pattern, sand rammers, trowel, leveler, vent road, molding sand mixture.

Material required: Molding sand, Aluminum alloy, clay, water, backing sand, parting sand, flux for cleaning of the molten metal

Schematic of the job:

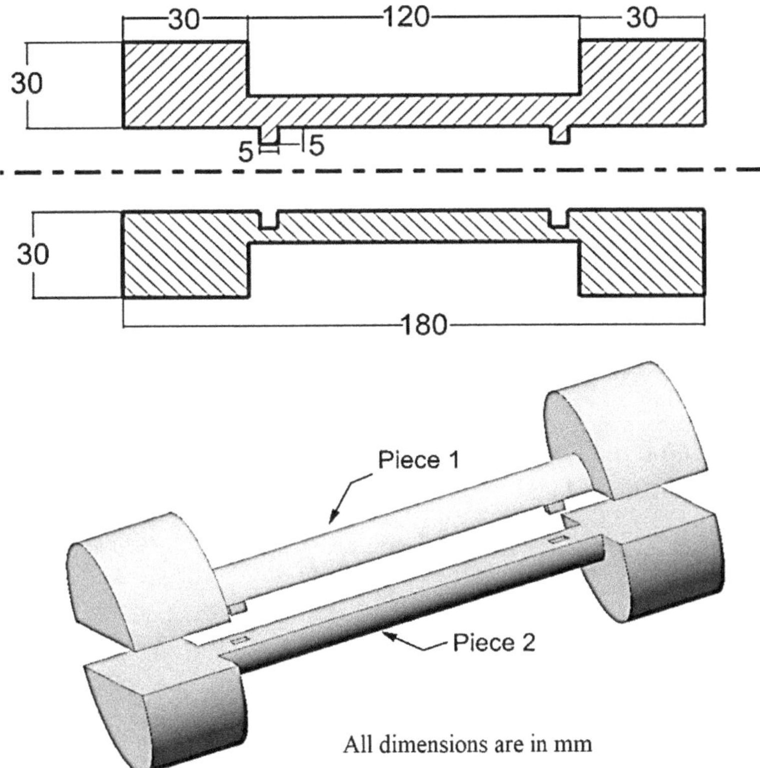

Figure P3.3 Split or two-piece pattern

Procedure:

    i.   Prepare the green sand by mixing water and clay with sand.

    ii.   Draw two diagonals to locate centre of the drag half.

    iii.   Place the drag half of the pattern on a bottom board (or a plane surface) keeping flat side of the pattern at the bottom. Also place the drag half of the molding box to surround the pattern (Figure P3.4 (a)).

    iv.   Sprinkle the parting sand over the pattern.

    v.   Fill the drag half of the molding box with the loose green sand around the pattern.

    vi.   Ram the sand packed mold with a rammer by first using the peen end followed by the butt end.

    vii. Strike-off bar is used to remove excess sand from the top surface of the drag half.

    viii. Turn the drag half upside down as shown in Figure P3.4 (b).

ix. Loose sand particles, if any, are blown off with the help of bellows.

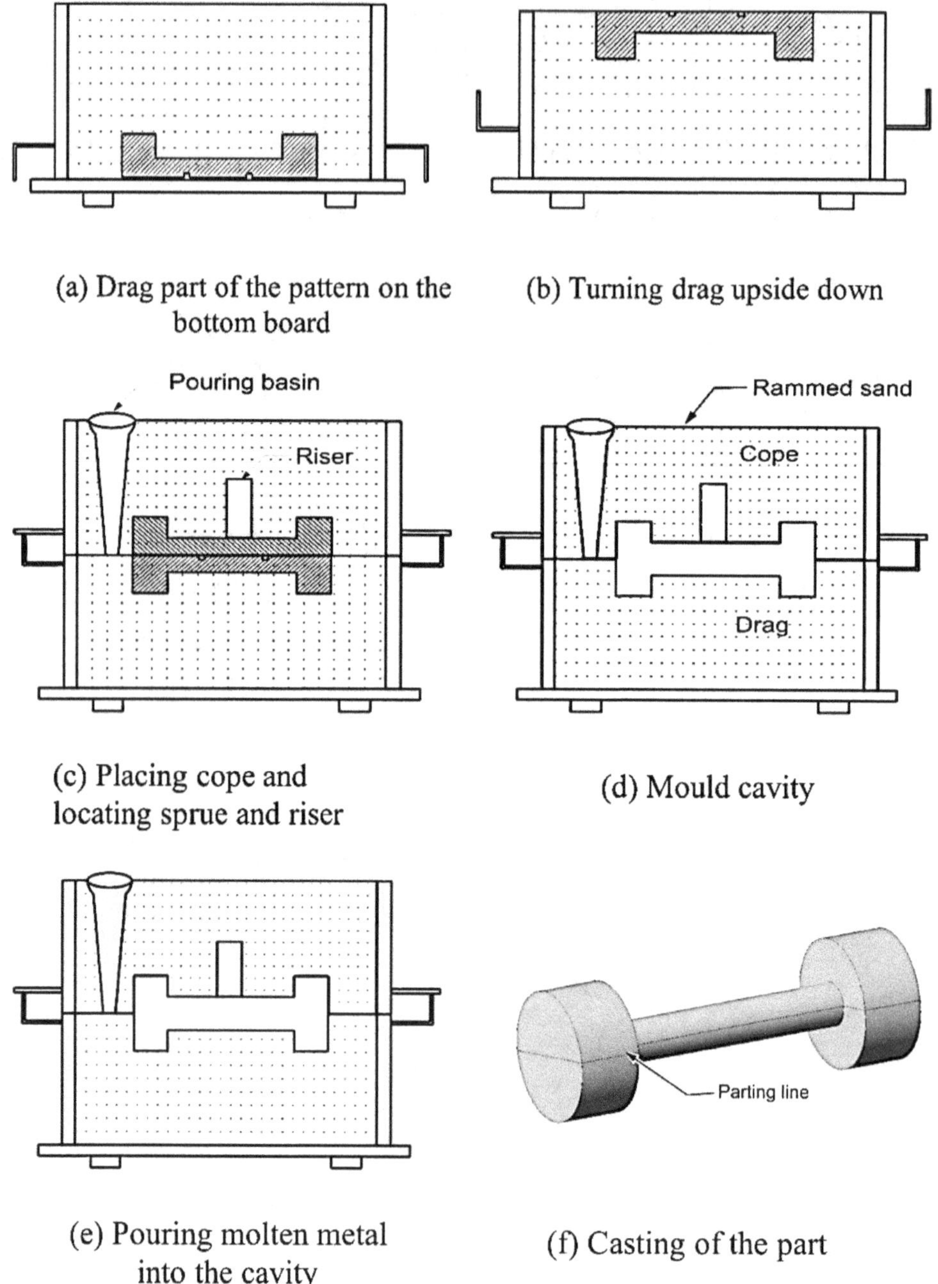

(a) Drag part of the pattern on the bottom board

(b) Turning drag upside down

(c) Placing cope and locating sprue and riser

(d) Mould cavity

(e) Pouring molten metal into the cavity

(f) Casting of the part

Figure P3.4 Mold preparation and casting using split pattern

x.   Place cope half of the pattern aligning it with the drag half. Completely fill the cope half of the molding box with the green molding sand.

xi. The riser pin is placed in such a manner to locate it on the highest point of the pattern. Sprue pin is also placed to touch the drag half and kept at about 5 to 6 cm from the pattern. (Figure P3.4 (c)).

xii. The upper surface of the drag half should be visible to you, sprinkle the parting sand on it.

xiii. The excessive sand seen on the top surface of cope is removed using a strike off bar.

xiv. Pierce vents which are thin holes through the top surface of the cope half with the help of a vent wire.

xv. The sprue pin and riser pin are removed from the rammed cope with care so that the mold is not damaged.

xvi. Remove the cope half lying on the drag half and place it inverted on a flat clean surface.

xvii. By using sand lifters, remove the extra sand particles from the drag half of the mold. Thereafter, using slick and lifter, make gates and a sprue base.

xviii. The two halves of the split pattern are removed from the mold very carefully by gently shaking it and using draw spikes and a mallet (Figure P3.4 (d)). Clean the cavities for any loose sand using bellows and place the cope half on the drag half and close the mold cavity.

xix. Repair the cavities wherever required.

xx. Let the mold dry and gain hardness under hot and dry environment.

xxi. Pour the molten metal into the mold cavity through the sprue opening (Figure P3.4 (e)).

xxii. Break the mold after solidification, take out and clean the casting (see Figure P3.4 (f)).

xxiii. The casting is visually inspected, and any surface and dimensional defects are identified.

xxiv. Fill in the workshop practice response sheet given at the end of chapter 2 and answer the questions provided therein. Get feedback of the instructor on the workpiece prepared by you.

Chapter 4

WELDING SHOP

4.1 INTRODUCTION

Welding process is about joining similar or dissimilar workpieces by the application of heat to obtain coalescence resulting in a strong joint between them. The application of heat may be done using either a direct method, such as an electric arc, or with the help of an indirect method, such as electrical resistance. Depending on the method of welding, pressure may also be applied. Furthermore, filler material, which helps fill the voids and compensate the loss of the material is also used. Welding is very commonly used in a variety of applications, such as pipelines, ships, farm equipment, automobiles, construction work, railways, and home appliances.

4.2 CLASSIFICATION OF WELDING PROCESSES

Several welding processes (Figure 4.1) are available which can be used for suitable applications. The welding processes can mainly be divided into two groups, pressure, and fusion, which are explained in the following paragraphs.

  i.   *Pressure welding:* In this process, application of pressure along with the heat is used to obtain the joining of two metals. Generally, the metal pieces are not heated to the extent of melting the metal but to bring them to the plastic stage and allow them to form a joint. Depending on how the heat is generated and the pressure is applied, several types of pressure welding processes are available, however, friction welding and resistance welding are the most common.
  ii.   *Fusion welding:* It is a process of joining the workpieces with the application of heat to bring their adjoining edges to the molten stage (or plastic stage) and allow for their solidification to get them joined. There is no application of force in this type of welding. A number of fusion welding processes are available, such as gas welding, electric arc, metal inert gas and tungsten inert gas welding.

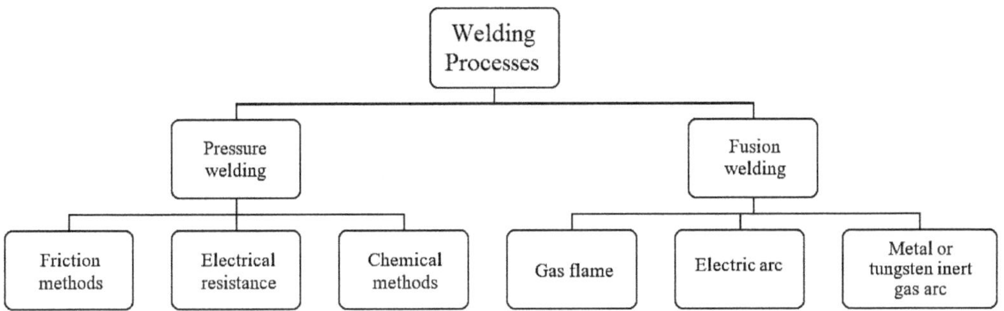

Figure 4.1 General classification of welding processes

A number of welding processes exist, which cannot be completely described here owing to the limited scope of this chapter. The classification of the welding process shown in Figure 4.1, is only a representative of a broader classification of the welding processes.

4.3 ELECTRIC ARC WELDING PROCEDURE

Electric arc welding method is one of the most important welding processes in which the metal workpieces are joined by melting the adjoining faces of the workpieces with the help of an electric arc. Following salient features of electric arc welding are worth noting.

i. The workpiece, which should be electrically conducting gets heat with the help of an arc. The arc is generated when a high current is passed through the electrode when it is brought near to the workpiece.

ii. The electrode is brought closer to the metal workpiece by keeping it at about 2 mm to 3 mm from the workpiece surface. The air gap between the electrode and the workpiece helps produce an electric arc due to ionization of the air available in the gap.

iii. The arc produces a very high temperature of about 4000°C, which is sufficient to melt high melting point alloys such as steel.

iv. The electrode is made to travel along the passage where the weld joint is required.

v. The electrode produces an arc as it continuously travels along adjoining faces of the workpieces to melt the metal followed by cooling of the molten metal and its solidification.

vi. Electrodes used in the arc welding process are coated with flux, which produces a gaseous shield around the molten metal. The gaseous shield prevents the reaction of the molten metal with the oxygen and nitrogen available in the atmosphere.

vii. The flux also has the chemical constituents which are helpful to remove impurities from the molten metal pool by forming a slag. This slag being of lower density comes to the top of the welded joint thereby protecting the molten metal from oxidation.

viii. The slag also helps slow down cooling of the weld joint. Slower cooling is considered better to get a strong and defect free joint.

ix. Lastly, the welded pieces are allowed to cool to the room temperature followed by chipping of slag formed on the upper surface of the weld to remove it. Figure 4.2 shows a schematic of the electric arc welding setup.

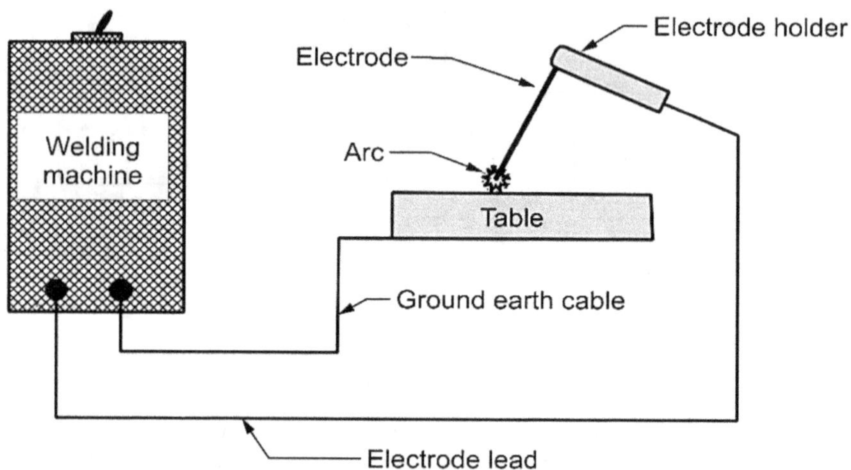

Figure 4.2 Schematic to show setup of electric arc welding

4.4 EQUIPMENT FOR ELECTRICAL ARC WELDING

As discussed in the previous section, the arc welding process needs establishing an electric circuit which comprises power supply, welding electrode, workpiece, and welding leads (or electric cables). Following paragraphs discuss important equipment required for the arc welding process.

The main task in the arc welding process is to complete the electric circuit to bring the electrode and the workpiece into the loop. The electrode serves as the filler rod also as it melts to supply the material to the welding joint area during the welding process. The following equipment are used for the electrical arc welding process.

i. *Power supply* provides either alternating current (AC) or direct current (DC) during the welding process. In case the power supply is AC (Figure 4.2), the direct power line is used along with a step-down transformer to reduce the voltage. The transformers may be either a single phase or a three phase.

   The DC power supply may either use straight polarity or reverse polarity. In straight polarity, (Figure 4.3 (a)) the electrode is connected to the negative terminal and the base metal to the positive terminal. The straight polarity helps in faster melting of the electrode and hence provides higher deposition rate due to the reason that flow of electrons is from the electrode to the workpiece. On the other hand, in reverse polarity, (Figure 4.3 (b)) an electrode is connected to the positive terminal and base metal to the negative terminal. The reverse polarity is useful for getting deeper penetration in the workpieces to be welded.

   Rectifiers are welding equipment for the welding process which are used for getting either AC or DC power supply. Welding machines ranging from 150 Ampere to 600 Ampere are easily available in the market; the current rating signifies the current available at the working terminal.

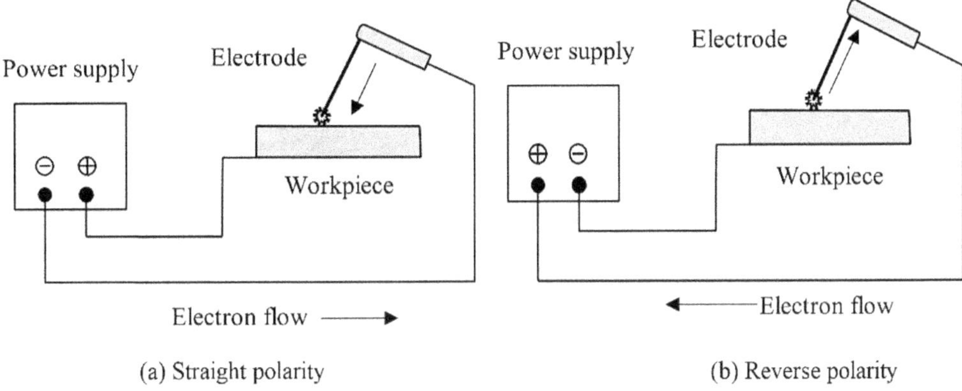

Figure 4.3 Welding polarities

ii.   *Welding cables* are required in the arc welding process primarily for performing two functions, (a) passing current from the power supply to the electrode holder as well as the electrode, and (b) making the electrical connection of the workpiece with the power supply unit for earthing.

iii.   The *cables* used for welding are made of copper and capable of carrying high current as per the requirement of the welding machine.

iv.   Arc welding electrode primarily caters to the following two functions, (a) providing filler metal to the welding joint for filling empty space between the joints, (b) supply flux, which helps form a protective environment during the welding process.

v.   *The electrode or filler rod* of arc welding process is made of similar material as that of the workpiece, which is also called the parent metal. An electrode is coated with the flux material throughout its length except the top end of about 20 mm length, which is kept bare to allow establishing electrical current between the electrode and the electrode holder. This is due to the reason that flux is an insulator of electricity. Figure 4.4 (a) shows important parts of an electrode.

vi.   *Electrode holder* is used to hold the electrode and connect it to the welding cables for establishing the electrical circuit. It also provides a handle for the user to securely hold the electrode for carrying out the welding process. An electrode holder is easy to handle, strong and remains at normal temperature during the welding process. It holds the electrode with the help of a spring mechanism. The user is prevented from an electric shock as insulation is provided to its handles. Figure 4.4 (b) shows a snapshot of an electrode holder.

vii.   *Ground clamp* is an important component used for grounding of the workpiece in the arc welding process. Therefore, a ground clamp helps by establishing an electrical connection between the worktable and the cable that leads to the earthing point. A ground clamp is strong enough to provide a low resistance electrical connection thereby preventing the user from electrical shock. Figure 4.4 (c) shows a typical ground clamp used in an arc welding machine.

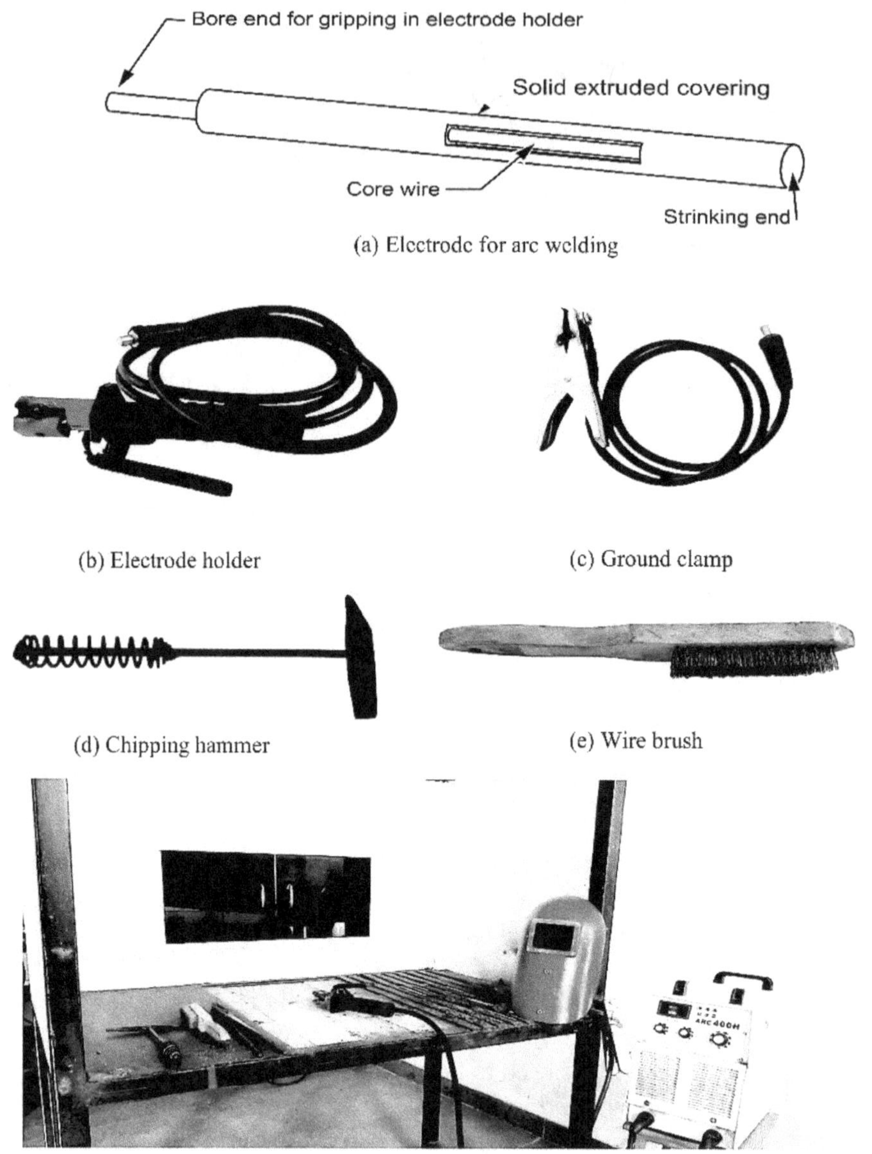

(a) Electrode for arc welding

(b) Electrode holder

(c) Ground clamp

(d) Chipping hammer

(e) Wire brush

(f) Welding, cabin, table, and other equipment

Figure 4.4 Equipment for arc welding

viii. *Wire brush* is required for cleaning the workpiece prior to its welding, whereas a chipping hammer is used to clean the welded job after the completion of the welding process. The chipping hammer has one of its ends as a sharp edge whereas the other one has a thick end. The sharp edge is used for the chipping of the spatter, which is due to small metal droplets that get deposited over the workpiece surface during the welding process. The thick end of the chipping hammer is used for removing the slag formation on welds. Figure 4.4 (d)-(e) shows the wire brush and the chipping hammer used in the welding shop.

ix.   *Welding table* is generally made of steel sheets (or plates) and structure with the primary purpose to properly align and position the parts to be welded. It is covered from the three sides to look like a cabin to ensure that other persons are not exposed to the welding arc even accidentally. The welding cabin or the space where the welding process is to be carried out should have sufficient provision for the exhaust of gases and fumes which if remain in the workplace become a health hazard. Figure 4.4 (f) shows a snapshot of the welding table and cabin.

4.5 SAFETY EQUIPMENT FOR WELDING

Several safety devices and equipment are required for the welding process, which are discussed here.

i.   *Face shield and welding goggles* are essential for protecting the eyes from the intense ultraviolet and infrared rays generated from the arc. These also help protect the welder from spatter (or flying particles) of hot metal that emerge during the welding process. A face shield is generally available in two variants, handheld and helmet type. The hand type face shield is made of light weight material and is convenient to use. Welding goggles are also made of light material and are more convenient to handle as both hands of the worker remain free for working when using them. Figure 4.5 (a)-(b) shows a typical face shield and goggles used during the arc welding process.

ii.   *Hand gloves* are also an important safety gear for welding, which help to protect a worker's hand from electric shocks and hot spatters. Figure 4.5 (c) shows an image of typical hand gloves used in welding.

Figure 4.5 Safety equipment for welding

4.6 WELDING JOINTS

The primary purpose of the welding process is to obtain a strong joint between the workpieces. There are a number of joint types available. Following types of welding joints are the most common, which are explained below and illustrated in Figure 4.6.

i.   *Butt joint* - is used to weld the edges of the two plates which are kept in the same plane. Depending upon the strength requirements, it can be given either a V type shape or a U type shape.

ii.   *Lap joint* - is used for joining the two overlapping plates in such a manner that the corner of each plate is joined with the surface of the other plate.

iii. *Edge joint* – is used for joining of the two parallel plates by welding them edge to edge.

iv.   *Tee joint* – as the name indicates, tee (or T) joint is used to weld two workpieces, which are at right angles to each other to form a T letter kind of shape.

v.   *Corner joint* – is made when edges of two workpieces are joined by keeping them at right angle to each other. Applications of corner joint are mainly found in frames and steel boxes.

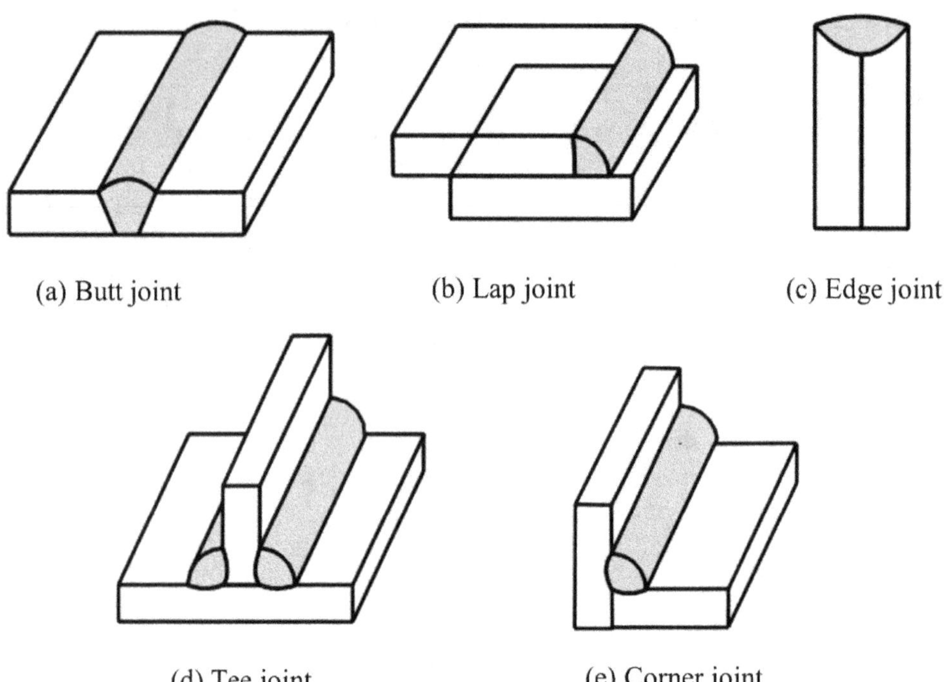

(a) Butt joint   (b) Lap joint   (c) Edge joint

(d) Tee joint   (e) Corner joint

Figure 4.6 Welding joints

4.7 WELDING DEFECTS

Knowledge of welding defects is essential to assess quality of the weld produced and take corrective action if required. A brief discussion of the welding defects is made in the following paragraphs. Few commonly found welding defects are shown in Figure 4.7.

i.   *Cracks* are generally formed in welds due to uneven heating and cooling of the workpiece and the weld pool. High amount of sulphur and carbon in the base metal also causes formation of cracks. A crack in a weld may be of microscopic size.

ii.   *Porosity* is a group of small sized holes formed on the surface of a weld. A major reason of porosity is absorption of gases, such as oxygen, nitrogen, and hydrogen in the molten weld pool. These gases which are absorbed in the molten metal are subsequently released during solidification thus forming pores in the weld.

iii.   *Poor fusion* is a defect which happens due to poor adhesion of the weld pool with the surface of the workpiece. This means that the workpiece and the weld are not properly joined leading to a poor-quality weld joint.

iv.   *Inclusions* are caused due to the presence of non-metallic foreign particles in the weld joint. This defect results in weakening of the joint.

v.   *Undercut* is also a welding defect which indicates that the weld is not properly filled and remains short to meet the top surface.

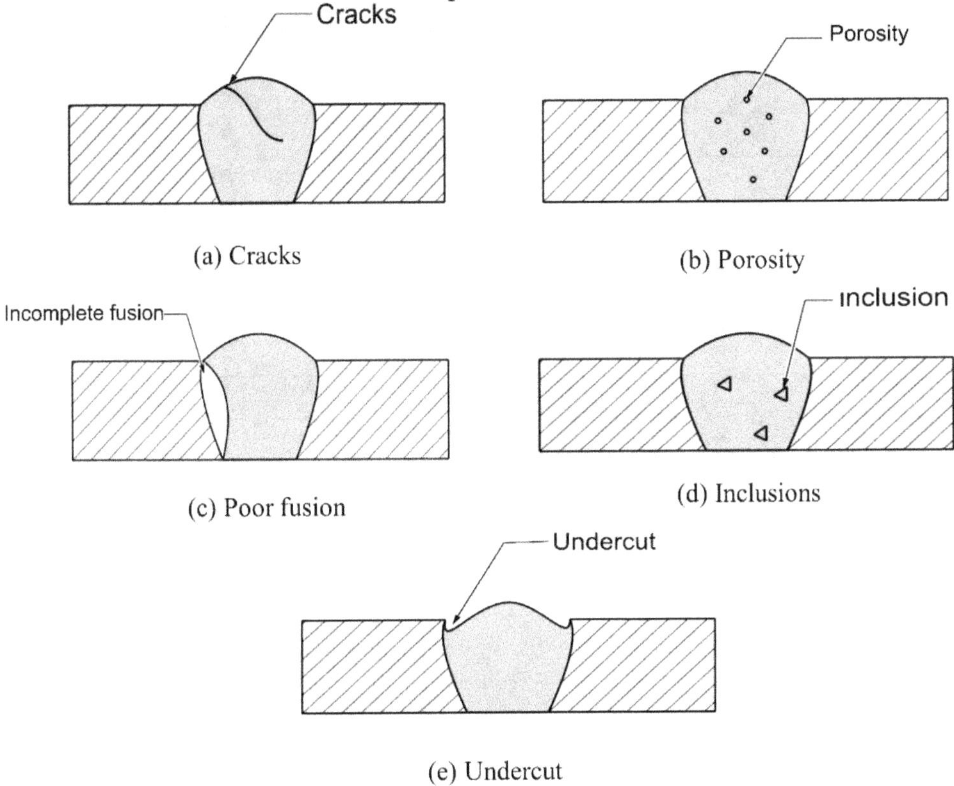

(a) Cracks

(b) Porosity

(c) Poor fusion

(d) Inclusions

(e) Undercut

Figure 4.7 Welding defects

4.8 PREPARING WORK FOR WELDING

Workpieces for making a weld joint should be properly cleaned to remove any foreign particles, dust, rust, and scale. The workpieces need to be aligned and kept stationary for carrying out the welding process. Thin workpieces can be welded as such; however, thick workpieces may need bevelling of their edges so that their edge surfaces get thoroughly fused to obtain a sound joint. Bevelling of edges of thick workpieces also helps in penetration of the metal from the filler rod to the weld joint area. Before starting the welding process, the following points must be taken care of.

- The welding cables are properly connected to the rectifier or the welding machine.
- An electrode of desired thickness and material compatible with the workpiece is available and inserted in the electrode holder in a proper manner.
- The welding current is set according to the size of the electrode.

Table 4.1 provides an idea of recommended process parameters, such as electrode diameter, current and voltage based on metal thickness, for arc welding of steel. The appropriate value of electrode diameter and process parameters should be selected based on the material of the workpieces and the recommendations provided by the electrode manufacturers.

Table 4.1 Selection of correct electrode diameter and other parameters

Metal thickness (mm)	Electrode diameter (mm)	Diameter (S.W.G)	Current (amp)	Voltage (V)
Up to 1.5	2.0	14	50-70	15
1.5-3	2.5	12	70-100	15
3-6	3.15	10	85-120	20
6-10	4.0	8	140-180	20
10-20	5.0	6	180-230	25
20-35	6.3	4	230-239	30

4.9 SAFETY PRECAUTIONS FOR WELDING

The welding process requires a few safety precautions to be followed, which are mentioned below.

i. Don't touch the just welded part or its surface after welding unless it cools down.

ii. Always keep a first aid kit near you in case your skin accidentally comes in contact with a hot surface.

iii. Keep combustible or inflammable materials away from the welding area as they may catch fire.

iv. Nobody should be allowed to stand near the operator to watch welding.

v. Before welding a container that may have been used for storing gasoline or other chemicals, make sure they are properly cleaned.

vi. Ensure that welding machine is properly grounded, and it's all leads are properly insulated.

vii. Always use a face shield or goggles for protecting eyes during welding.

viii. Welding cables should be prevented from coming in contact with the hot metal or liquids, such as water, oil, and grease.

ix. Protective equipment such as hand gloves, leather shoes and apron should be always be worn before starting the welding process.

x. Turn off the welding machine when not in use.

xi. Always keep yourself insulated during the welding process.

PRACTICE NO. 4.1

Job: Prepare a butt joint in flat position the electric arc welding process.

Objectives: Learn by practice various operations related to arc welding and their respective tool handling for making a butt weld joint and identify common welding defects.

Machines, equipment, and tools required: Arc welding machine, mild steel electrode, electrode holder, ground clamp, tong, face shield, apron, hand gloves, welding table, bench vice, rough flat file, try square, steel rule, wire brush, ball peen hammer, chipping hammer.

Material required: Two mild steel workpieces, each of 50 mm x 30 mm x 5 mm size.

Schematic of the job:

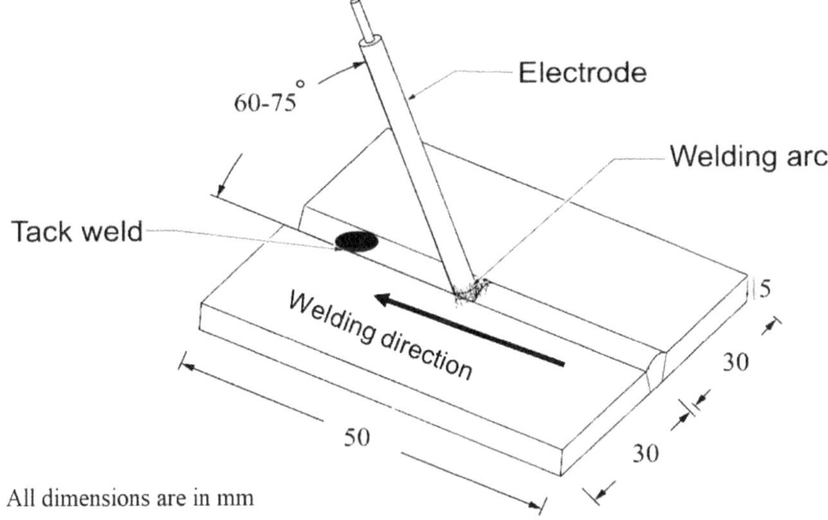

Figure P4.1 Butt joint

Procedure:

i. Two mild steel workpieces are taken and sized to the dimension shown in Figure P4.1. The workpieces are cleaned to remove any dust, oil, or grease.

ii. Sharp corners or burrs, if any, on the workpieces are removed by filing.

iii. Position the workpieces in such a way that they are close enough for welding leaving a very thin space in between them for better penetration of the molten metal for making the weld joints.

iv. Give electric power supply to the welding set.

v. Connect cables, one with the electrode holder and the other with the earthing wire.

vi. The electrode is placed in the electrode holder and the welding current is set referring to Table 1.1 and recommendations of the electrode manufacturer.

vii. The earth (or ground) clamp is fastened to the welding table. The welding machine is switched ON.

viii. Wear safety gears, such as apron, face shield and struck the arc. Ensure that all safety precautions are strictly followed.

ix. Struck the arc to tack weld the workpieces to ensure they don't move during welding process. It is to be noted that tack weld is a small weld spot which helps in tying the workpieces together but is not strong enough. This is followed by making the first run of the welding along the specified path to fill the root gap.

x.   Make second run of the welding process ensuring uniform movement and proper weaving to make the weld joint. During the welding process, keep angle of the electrode to be 60° to 75° with the plate surface.

xi.   Allow the weld to cool and remove the slag formed on its top surface.

xii. Make use of a file to remove the spatter around the weld joint using filing.

xiii. Fill in the workshop practice response sheet given at the end of chapter 2 and answer the questions provided therein. Get feedback of the instructor on the workpiece prepared by you.

PRACTICE NO. 4.2

Job: Prepare a Tee joint in flat position by electric arc welding.

Objectives: Learn by practice the operations and procedures needed for making a Tee joint and identify common welding defects.

Machines, equipment, and tools required: Arc welding machine, mild steel electrode, electrode holder, ground clamp, tong, face shield, apron, hand gloves, welding table, bench vice, rough flat file, try square, steel rule, wire brush, ball peen hammer, chipping hammer.

Material required: Two mild steel workpieces of 50 mm x 30 mm x 5 mm each.

Schematic of the job:

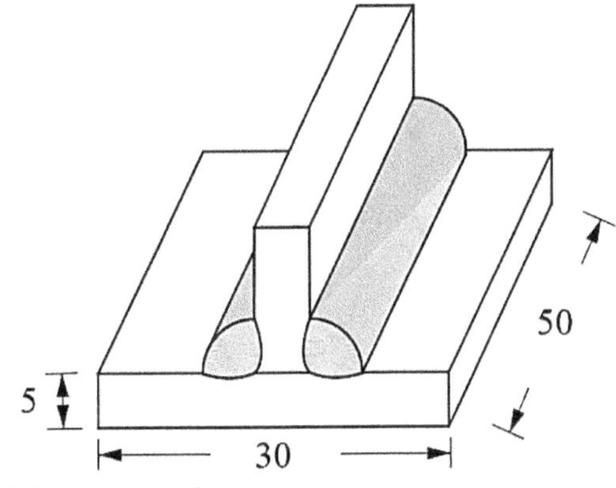

Figure P4.2 Tee Joint

Procedure:

i. Take two mild steel workpieces of given dimensions and clean their surfaces thoroughly from rust, dust particles, oil and grease.

ii. Remove the sharp corners and burrs found on the workpieces by filing or grinding.

iii. The workpieces are positioned on the welding table so as to form a T shape (Figure P4.2).

iv. A compatible electrode is fitted into the electrode holder and the welding current is set referring to Table 1.1 and recommendations of the electrode manufacturer.

v. The ground clamp is fastened to the welding table.

vi. Wear all the safety gears, such as apron, face shield and struck the arc. Ensure that all safety precautions are strictly followed.

vii. The arc is struck, and the workpieces are tack-welded at both the ends.

viii. The alignment of the T joint is checked, and the tack-welded workpieces are reset, if required.

ix. Welding is carried out throughout the length of the T joint as shown in figure P4.2.

x. Remove the slag and spatter from the workpiece with the help of a chipping hammer to clean the joint.

xi. Fill in the workshop practice response sheet given at the end of chapter 2 and answer the questions provided therein. Get feedback of the instructor on the workpiece prepared by you.

PRACTICE NO. 4.3

Job: Make a corner joint in vertical position by electric arc welding

Objectives: Learn by practice arc welding procedures and operations for making a corner weld joint and identify the welding defects.

Machines, equipment, and tools required: Arc welding machine, mild steel electrodes, electrode holder, ground clamp, flat nose tong, face shield, apron, hand gloves, metallic worktable, bench vice, rough flat file, try square, steel rule, wire brush, ball peen hammer, chipping hammer, chisel and grinding machine.

Material required: Two mild steel pieces, each of 50 mm x 30 mm x 5 mm.

Schematic of the job:

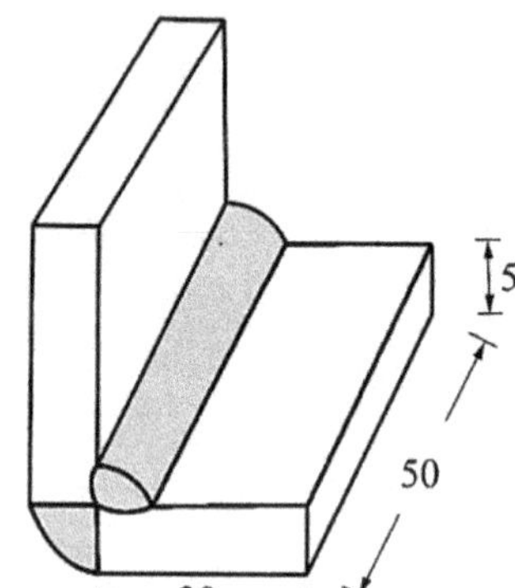

Figure P4.3 Corner joint in vertical position

Procedure:

i.   Take the two mild steel workpieces of given dimensions and clean their surfaces thoroughly from rust, scale, dust particles, oil, and grease.

ii.   The two workpieces are positioned on the welding table to form letter 'L' like shape making use of tongs (Figure P4.3).

iii.   The electrode is fitted in the electrode holder and the welding current is set referring to Table 1.1 and recommendations of the electrode manufacturer.

iv.   The ground clamp is fastened to the welding table.

v.   Wear safety gears, such as apron, face shield and struck the arc. Ensure that all stated precautions are strictly followed.

vi.   The workpieces are tack welded at both the ends as well as at the center of the joint.

vii.   The alignment of the corner joint is checked. Thereafter, the welding is done for complete length of the workpiece from both the front and the back side of the corner joint.

viii.   The slag formed on the weld is removed by using the chipping hammer.

ix.   Filing is done to remove any spatter around the weld.

x.   Fill in the workshop practice response sheet given at the end of chapter 2 and answer the questions provided therein. Get feedback of the instructor on the workpiece prepared by you.

PRACTICE NO. 4.4

Job: Prepare lap joint in horizontal position using the electric arc welding process.

Objectives: Learn by practice the arc welding procedure and operation for making a lap weld joint and identify the welding defects.

Machines, equipment, and tools required: Arc welding machine, mild steel electrodes, electrode holder, ground clamp, flat nose tong, face shield, apron, hand gloves, metallic worktable, bench vice, rough flat file, try square, steel rule, wire brush, ball peen hammer, chipping hammer, chisel and grinding machine.

Material required: Two mild steel workpieces of 50 mm x 30 mm x 5 mm each.

Schematic of the job:

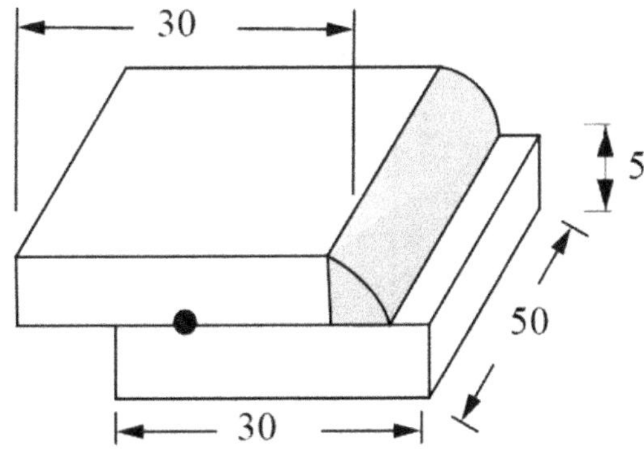

Figure P4.4 Lap weld joint in horizontal position

Procedure:

i. Take the two mild steel workpieces of given dimensions and clean their surfaces thoroughly from rust, scale, dust particles, oil, and grease.

ii. The two workpieces are placed on the welding table in overlapping position with each other with the help of tongs (see Figure P4.4).

iii. A mild steel electrode is fitted in the electrode holder and the welding current is set referring to Table 1.1 and recommendations of the electrode manufacturer.

iv. The ground clamp is fastened to the welding table.

v. Wear all the safety gears, such as apron, face shield and struck the arc. Ensure that all the stated precautions are strictly followed.

vi. The alignment of the lap joint is checked.

vii. The slag formation on the weld surface is removed with the help of a chipping hammer. The spatter is also cleaned.

viii. Fill in the workshop practice response sheet given at the end of chapter 2 and answer the questions provided therein. Get feedback of the instructor on the workpiece prepared by you.

PRACTICE NO. 4.5

Job: Make an edge joint in the vertical position by the electric arc welding process.

Objectives: Learn by practice welding operations and procedures for making an edge weld joint and identify the welding defects.

Machines, equipment, and tools required: Arc welding machine, mild steel electrodes, electrode holder, ground clamp, flat nose tong, face shield, apron, hand gloves, metallic worktable, bench vice, rough flat file, try square, steel rule, wire brush, ball peen hammer, chipping hammer, chisel and grinding machine.

Material required: Two mild steel pieces of 50 mm x 30 mm x 5 mm each.

Schematic of the job:

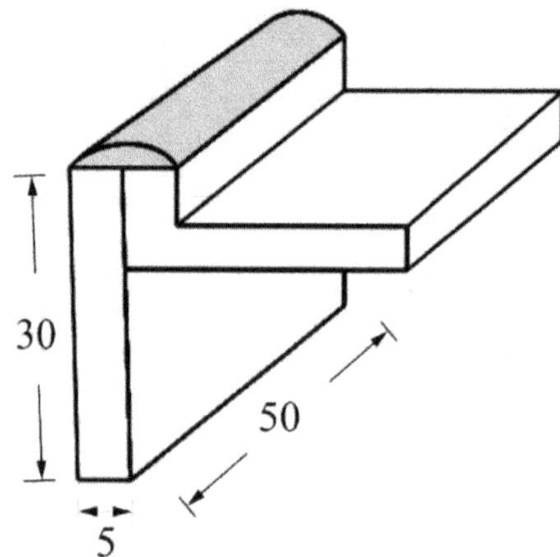

All dimensions are in mm

Figure P4.5 Edge joint in a vertical position

Procedure:

i. Take the two mild steel workpieces of given dimensions and clean the surfaces thoroughly from rust, dust particles, oil, and grease.

ii. The two workpieces are placed on the welding table such that one workpiece is placed vertically on the edge of the other workpiece (Figure P4.5). The tongs are used for holding the workpieces for this purpose.

iii. The mild steel electrode is fitted in the electrode holder and the welding current is set as per recommendations given in Table 1.1 and the electrode manufacturer.

iv. The ground clamp is fastened to the welding table.

v. Wearing the apron and using the face shield, an arc is struck, and the workpieces are tack welded at both the ends as well as at the mid position of the joint.

vi. The alignment of the edge joint is checked.

vii. Welding is completed for complete length of the workpieces.

ix. The slag formation on the welds is removed with the help of a chipping hammer. The spatter is also cleaned.

viii. Fill in the workshop practice response sheet given at the end of chapter 2 and answer the questions provided therein. Get feedback of the instructor on the workpiece prepared by you.

Chapter 5

FITTING SHOP

5.1 INTRODUCTION

Machine tools, such as lathe are used for material removal at a fast pace. However, many times there is a requirement to do minor fitting and repairing operations manually, with the help of hand tools. Such manual operations which are done primarily for mating of parts are called fitting operations. Sometimes parts are made entirely with the help of fitting operations by holding them in the bench vice, which is also called 'bench work'. Accuracy of the work done in the fitting shop depends on the fitter's experience and skill. Both the bench work and the fitting tools, need hand tools and human effort. Examples of fitting operations are filing, chipping, scraping, sawing, drilling, and tapping.

5.2 FITTING TOOLS

A number of tools are used in the fitting shop, which are classified into holding, marking, measuring, cutting, finishing and striking tools. Figure 5.1 shows classification of the fitting tools whereas the following paragraphs provide their description.

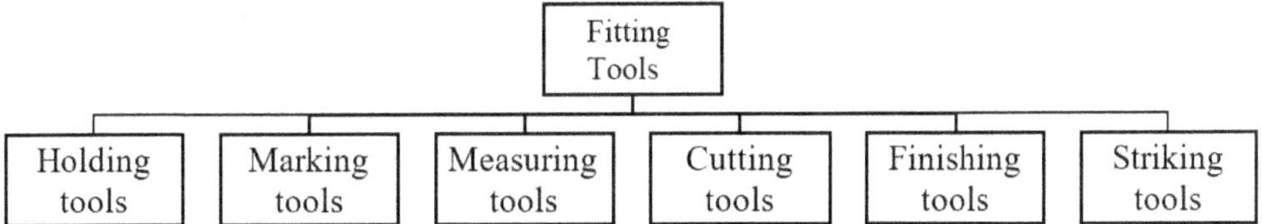

Figure 5.1 Tools used in a fitting shop

5.2.1 Holding tools

Holding tools are used in the fitting shop for holding or tightly fixing a job so that the fitting operations can be carried out on it. These tools are also referred to as work holding tools or devices.

i. *Workbench* (Figure 5.2 (a)) is used for placing various components on it as well as for securely placing a bench vice. The bench vice, which is discussed later in this section, is used to hold the workpieces. Furthermore, the workbench also provides support to the worker while performing fitting tasks.

ii. *Bench Vice* (Figure 5.2 (b)) is used for holding specimens or workpieces between its two jaws. It is mounted on the workbench with the help of nuts and bolts. A bench vice has two major parts used for holding: a fixed jaw and a movable jaw. The size of a vice is defined by length of its jaws. The body of the bench vice is generally made of cast iron, which has high modulus of elasticity but low impact strength. Due to low impact strength of the bench vice, it is a good idea to avoid hammering of the job held on it.

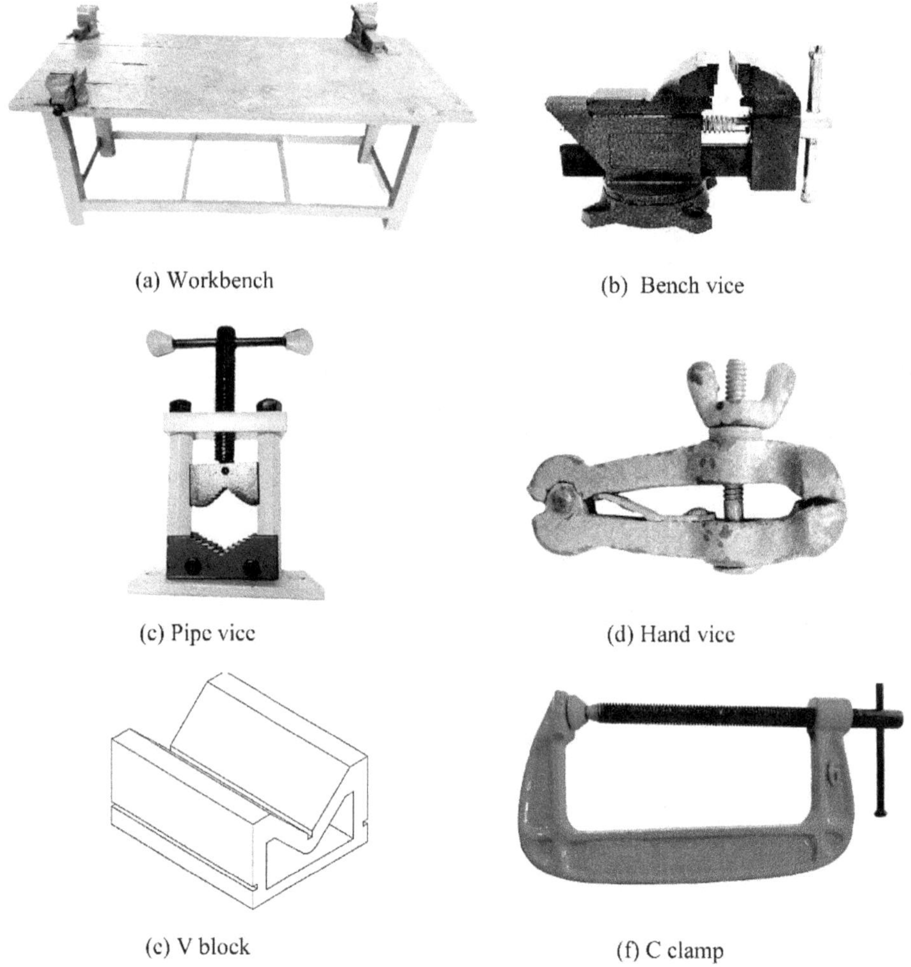

(a) Workbench

(b) Bench vice

(c) Pipe vice

(d) Hand vice

(c) V block

(f) C clamp

Figure 5.2 Holding tools

iii. *Pipe vice* (Figure 5.2 (c)) is identical to bench vice, except that its jaws are designed to hold round shape jobs. The jaws of this vice are not flat but have two surfaces, which are at a right angle.

iv.   *Hand vice* (Figure 5.2 (d)) is a portable device which is also used for holding workpieces in between its jaws. A hand vice is also used for holding jobs which cannot be brought to a bench vice.

v.   *V-block* (Figure 5.2 (e)) is a metallic, rectangular, or square shaped block with a V-groove at its two opposite sides. The included angle of the V groove is usually 90 degrees. A V – block can be used to clamp round shaped parts with the help of a clamp.

vi.   *C- Clamp* (Figure 5.2 (f)) is used for holding two workpieces together. It has a fixed jaw which is shaped like letter 'C', whereas its second jaw is of round shape, is movable and attached to the end of a screw. Principle of the C-clamp is like the bench vice except that it is portable and not fixed on a bench. C-clamp is also used for holding of workpieces in a V block.

5.2.2 Marking tools

The primary function of a marking tool in a fitting shop is to mark the surface of a workpiece, for which scriber and punches are the most used tools in the fitting shop.

i.   *Scriber* (Figure 5.3(a)) is a thin steel tool, which is hardened and tempered. It is used to scribe or mark lines on a workpiece. The scribers are available in different lengths ranging from 125 mm to 250 mm. A scriber has two sharp pointed edges, one of which is straight and the other one is bent. The bent end is used to mark lines in those areas which are inaccessible by the straight edge.

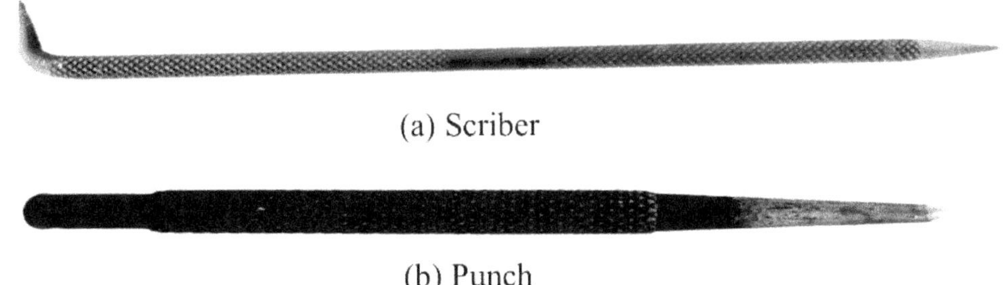

(a) Scriber

(b) Punch

Figure 5.3 Marking tools

ii.   *Punch* (Figure 5.3 (b)) is made of high carbon steel and is used to make noticeable indentations on the workpiece. A punch is usually specified with the help of its length and diameter. One end of the punch is tapered and conical in shape with a pointed hardened edge. The punches are generally of two types, dot punch and center punch. As the name appears, the dot punch is used for making a dot as its conical point is hardened with an angle of 60°. The center punch is used to mark a center for carrying out operations like drilling to make holes. The included angle of the pointed conical edge of the center punch is 90°.

5.2.3 Measuring tools

Measuring tools for fitting shop are essential to cater to several tasks such as length and angle measurement, which are discussed in the following paragraphs.

i.   *Steel rule* (Figure 5.4 (a)) is a geometric instrument used to measure the length in metric and inch system.

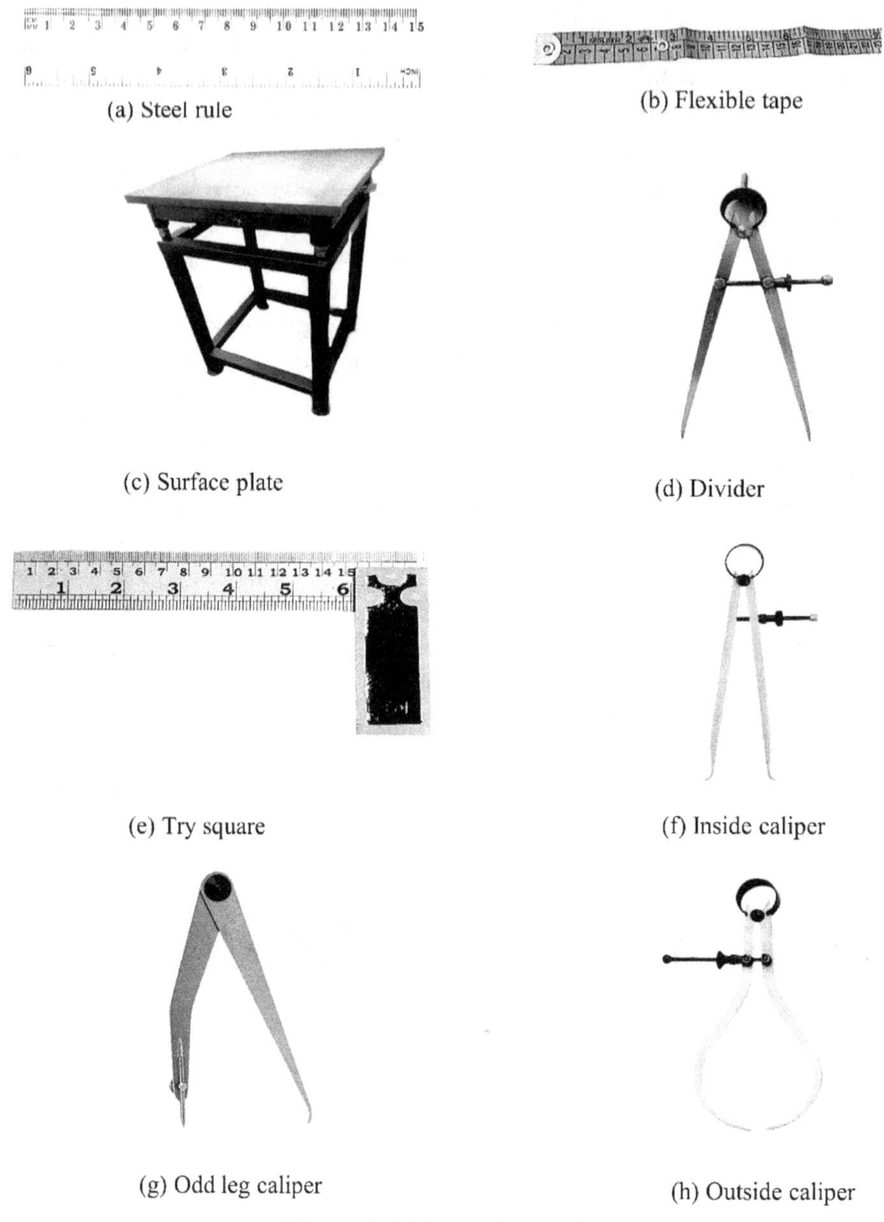

(a) Steel rule

(b) Flexible tape

(c) Surface plate

(d) Divider

(e) Try square

(f) Inside caliper

(g) Odd leg caliper

(h) Outside caliper

Figure 5.4 Measuring tools

ii.   *Flexible tape* (Figure 5.4 (b)) has markings to measure the length in millimeters, centimeters, and inches just like a steel foot rule.

iii. *Surface plate* (Figure 5.4 (c)) provides an accurately grounded flat surface to place a workpiece for inspecting and marking a specimen with precision and for layout and tooling setup.

iv.   *Divider* (Figure 5.4 (d)) is an instrument used for marking circular arcs, marking perpendicular lines, bisecting lines and other similar tasks on workpieces.

v.   *Try square* (Figure 5.4 (e)) is useful for measuring right angles between surfaces. It comprises a thin hardened steel blade and a thick cast iron base, the included angle between the two is 90°. Try squares of different sizes are available in the market.

vi.   *Inside caliper* (Figure 5.4 (f)) is used for measuring inner size of a workpiece's feature like hole. It has two non-inverted legs, which are connected with the help of a screw and nut mechanism.

vii.   *Odd leg caliper* (Figure 5.4 (g)) is used for marking parallel lines and locating the center from the finished edge.

viii.   *Outside caliper* (Figure 5.4 (h)) can be used for measuring an object's external dimensions. It has two inverted legs, which are connected using a screw and nut mechanism.

5.2.4 Precision measuring tools

This section discusses three precision measuring tools, namely vernier caliper, vernier height gauge and micrometer. However, several other precision measuring tools also make use of similar principles. The least count of a precision measuring tool is an important parameter, which refers to the smallest and most accurate value that can be measured with the help of the measuring tool. Equation 5.1 is used to find out the least count.

$$Least\ count = \frac{smallest\ divison\ on\ the\ main\ scale}{Total\ number\ of\ divisons\ on\ the\ secondary\ scale} \quad (5.1)$$

A discussion on precision measuring tools is discussed in the following paragraphs.

i.   *Vernier calipers* accurately measure linear dimensions of a workpiece (figure 5.5 (b)), whether outside or inside. It also has a provision to measure depth. It has two jaws, one of which is part of vernier scale, and the other one is fixed at one end of its main scale. For example, if in a vernier caliper, the number of divisions in the vernier scale is 50 and the smallest main scale division is 1 mm, the least count will be 1/50 or 0.02-mm. Vernier calipers with different least counts ranging from 0.1 mm to 0.02 mm are normally available in the market. Figure 5.5 (a) shows the schematic of a vernier caliper.

ii.   *Micrometer* is used to measure the diameter of cylindrical shaped workpieces. It has two scales, a main scale, and a secondary scale. The secondary scale has markings on cylindrical surface of a screw, which moves on a nut aligned with the main scale. The least count of a micrometer is calculated with the help of the smallest division on the main scale and the number of secondary scale divisions using the formula given in equation 5.1. Micrometers with different least counts are available in the market, the most common being a micrometer having main scale division of 1 mm and 100 divisions in the secondary scale, which provides the least count of 0.01 mm.

iii.   *Vernier height gauge* is used to measure vertical distances and heights of holes. It works on the same principle as that of the vernier caliper, i.e., a main scale and a vernier scale. The range of maximum height that can be measured with a vernier height gauge generally varies from 150 mm to 1000 mm. Figure 5.5 (c) illustrates a vernier height gauge.

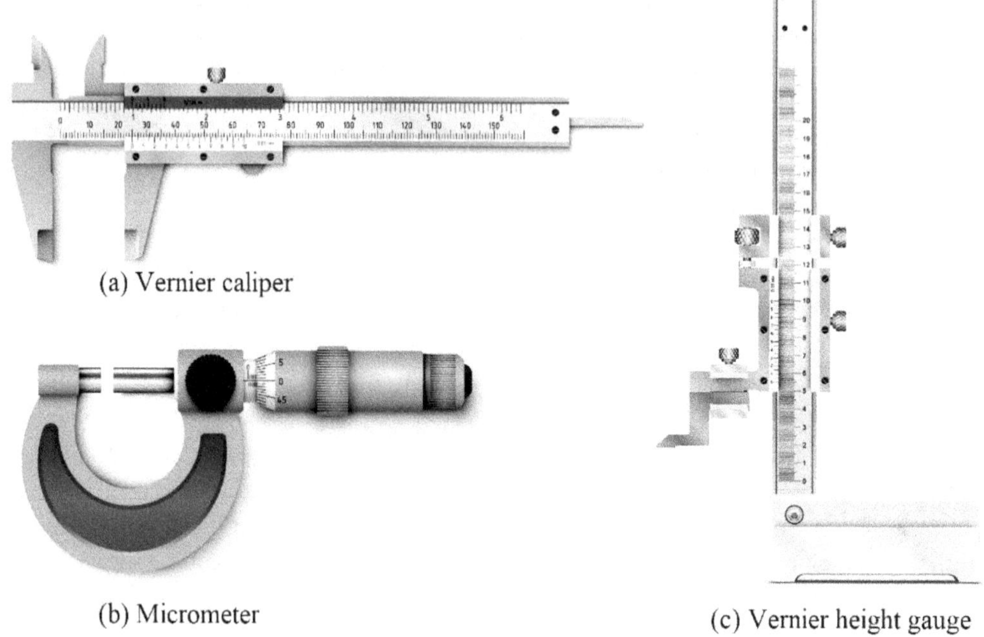

Figure 5.5 Precision measuring tools

5.2.5 Cutting tools

A number of cutting tools are used in the fitting shop to carry out a variety of operations, which are described in the following paragraphs.

i.   *Hacksaw* (Figure 5.6 (a)) is used to cut workpieces, such as rods, bars, sheets, and pipes. It has a frame made of mild steel, a handle, and two prongs to tie ends of the hacksaw blade. Depending on the material of the blade and specifications of its teeth, which is described in number of teeth per unit length in inches, several blade types are available. Generally, blades are classified as coarse (8-14 teeth per inch), medium (16-20 teeth per inch), and fine (24-32 teeth per inch).

ii.   *Chisels* (Figure 5.6 (b)) are used to remove excess material from a workpiece and cutting off thin sheets. Chisels are made of carbon steel and cross section of a chisel is generally shaped hexagonal or octagonal. Chisels are heat treated to make a tough body whereas its cutting edge is hardened and tempered. Further, annealing heat treatment is also used to minimize the stresses induced in a chisel's body. Generally, included angle of a chisel's cutting edge is about 60°.

iii. *Twist drills* (Figure 5.6 (c)) are made of high-speed steel and used to create holes. Straight and taper shank twist drills are most used.

iv.   *Taps* (Figure 5.6 (d)) are used to cut internal threads in the holes which are already drilled. A hand tap set generally has three taps, namely a taper tap, an intermediate tap and a plug (or bottoming) tap, which are used in sequence to create an internal thread of a particular size. Each set includes a taper tap, intermediate tap and plug or bottoming tap, which are used in sequence to create an internal thread of a particular size. Taps are generally made of high carbon steel or high-speed steel. Figure 5.6 (d) shows a tap wrench used in a typical fitting shop.

v.   *Dies* (Figure 5.6 (e)) are used to cut external thread. A die may be either of solid type or split type. A die stock is used for holding the die and adjusting the die gap to carry out the threading operation. Generally, dies are made of medium carbon or high carbon steel.

Figure 5.6 Cutting tools

vi. *Bench drill* (Figure 5.6(f)) is a powered machine tool which is used to drill holes. Bench drill is the most widely used machine in a fitting shop. Twist drills are attached to the main spindle of a bench drill with the help of an attachment like a collet or drill chuck for machining of holes. The following procedure is adopted for performing drilling work on a bench drill.

    a)   Select a correct sized twist drill suitable for the job, put it into the chuck or collet and lock it firmly.

    b)   Adjust the speed of the bench drill's spindle using the mechanism provided on the machine. High speed is to be used for small diameter holes and soft materials, whereas low speed is required for large diameter holes and hard materials.

c)   With the help of a center punch, mark the workpiece at positions where the holes are to be drilled.

d)   Hold the workpiece firmly in the vice on the bench drill's table by clamping it directly onto the machine table.

e)   Put on the power, locate the punch mark, and apply appropriate pressure on the rotating drill using feed handle of the bench drill.

f)   Continue to apply pressure to complete the drilling operation. Cutting oil may be used for cooling on steel workpieces at the drilling point.

g)   After drilling operation is complete, the pressure is slightly released.

h)   Slowly retract the drill out of the work and switch-off the power supply.

5.2.6 Finishing tools

Finishing tools are used for removing a thin layer of material from the workpiece's surface primarily to improve surface finish. Important finishing tools used in the fitting shop are mentioned in the following paragraphs.

i.   *Reamer* is used for carrying out reaming operation which refers to sizing and finishing of a drilled hole. Reaming operation needs a drilled hole which has a diameter smaller than the required hole size as the hole gets slightly enlarged in diameter with the reaming operation. The reaming operation can be done with the help of either a bench drill or a tap wrench. The reamer's body has a slight taper at its working end to facilitate easy entry into the hole during operation. The reamer is rotated in clockwise direction both during machining and retracing. Figure 5.7 shows reamers used for sizing and finishing a drilled hole.

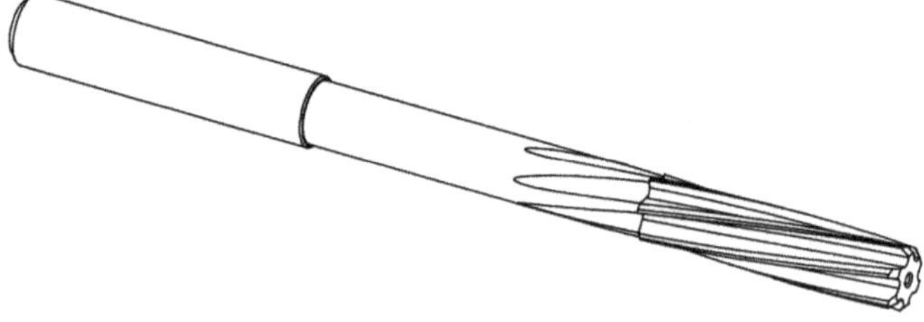

Figure 5.7 Reamer

ii.   *Files* are used to remove the extra material from the workpiece surface to get high surface finish and correct size dimensions. Files are multi-point cutting tools, which are used to remove the material by rubbing on the workpiece. Files come in a variety of sizes, shapes, and degrees of coarseness. Generally, three types of files are available: course, medium and fine. Coarse files are used to remove the material at faster pace whereas the fine files remove material at lower pace and better finish. The filing operations may be classified into two types, as mentioned below.

a)   *Cross filing:* In this method, filing is done in a direction perpendicular to the workpiece's axis (Figure 5.8 (a))

b)   *Draw filing:* In this method, filing is done in a direction parallel to the workpiece's axis (Figure 5.8 (b))

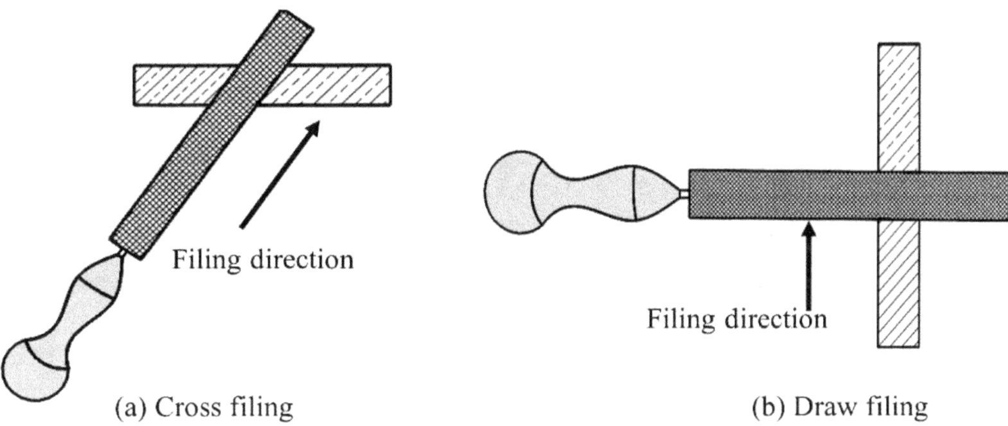

Figure 5.8 Methods of filing

Further files are classified based on their shape, which is described in the following paragraphs with their illustrations provided in Figure 5.9.

- *Flat file* (Figure 5.9 (a)) is having a rectangular cross-section and it is used for removing the material from flat surfaces.
- *Square file* (Figure 5.9 (b)) is helpful to remove the material from the inside corners of a workpiece which are at right angles. This is due to the reason that flat files may find such faces to be inaccessible.

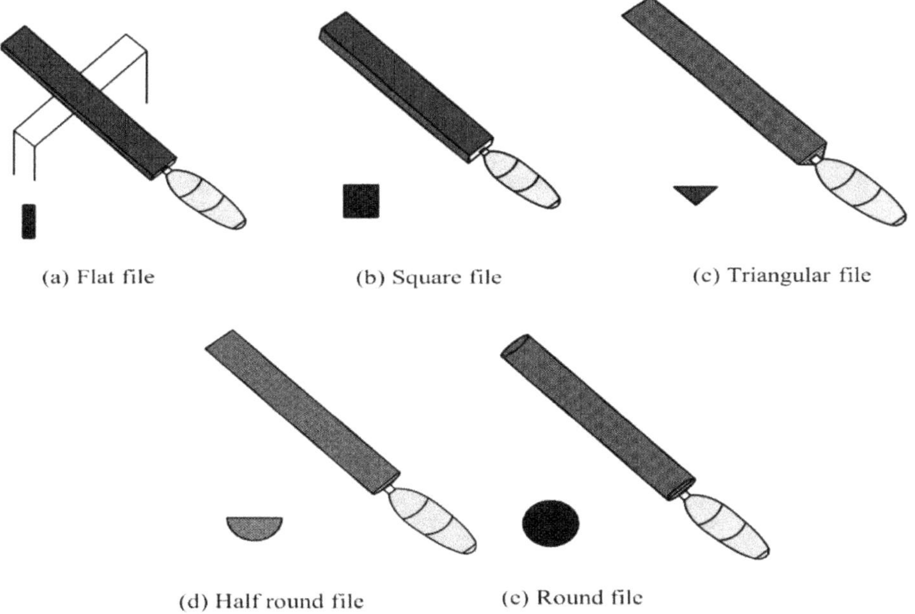

Figure 5.9 File types according to their cross-section shape

- *Triangular file* (Figure 5.9 (c)) has a triangular cross-section and is used to remove material from the inclined faces and such areas where flat or square files cannot reach.

- *Round file* (Figure 5.9 (d)) has a circular cross-section. It removes the material from the cylindrical and curved surfaces of workpieces to achieve good surface finish.

- *Half-round file* (Figure 5.9 (e)) has a semi-circular cross section, which means a flat face and a round face. It is used to file such faces which are a combination of round and flat faces.

- *Swiss or Needle file* has a small area of cross section, which makes it possible to access narrow areas like keyholes for filing operations.

5.2.7 Striking tools

Striking tools are used for forcefully hitting and striking the workpiece to help perform the intended operation. Normally, three types of striking tools are used in the fitting shop, namely cross-peen hammer, claw hammer and mallet, which are explained in the following paragraphs and their illustrations are provided in Figure 5.10

i. *Mallet* (Figure 5.10 (a)) is used for striking the chisel on a wooden component. It is made up of hardwood and may be of either round or rectangular shape.

ii. *Cross peen hammer* (Figure 5.10 (b)) comprises a steel body and a handle made of wood or plastic. The body of a cross-peen hammer has two parts, a narrow edge peen which is a right angle to the handle and an opposite face with a flat surface.

iii. *Claw hammer* (Figure 5.10 (c)) also has a steel body and a handle made of wood or plastic. It is used for striking as well as for pulling nails from the wood. The claw face is used for pulling out the nails, whereas the head face is used to drive the nails.

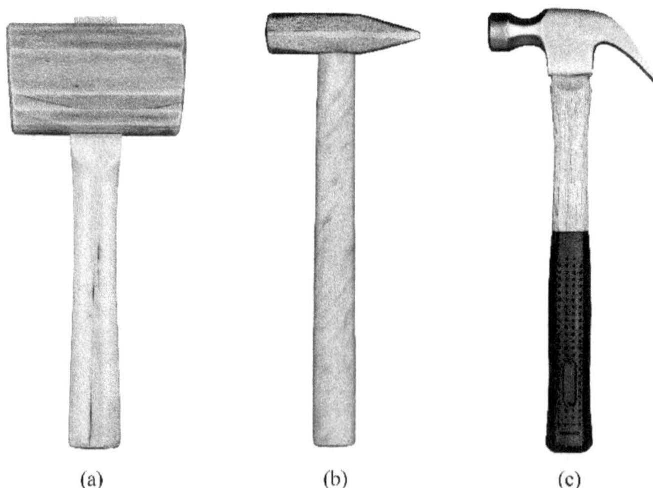

Figure 5.10 Striking tools: (a) Mallet hammer, (b) cross peen hammer, and (c) Claw hammer

5.3 FITTING OPERATIONS

Several operations are used in the fitting shop, which are mentioned below.

i. Filing
ii. Measuring
iii. Marking
iv. Punching
v. Sawing
vi. Drilling
vii. Reaming
viii. Tapping

5.4 SAFETY AND BEST PRACTICES

Carrying out bench working and fitting operations in a fitting shop needs knowledge of safe and correct work practices, which are discussed in the following paragraphs. Further, the best practices to be followed are illustrated in Figure 5.11 as the right and wrong ways of carrying out a fitting operation.

i. Hands and tools should be kept clean from dirt, grease, and oil by wiping them out. It is always safe to use dry tools as an uncleaned tool is dangerous due to risk of slipping out of hands.

ii. Sharp tools should not be kept in pockets.

iii. Working on the workpiece should be done either on the table or by securely placing it in the vice. The workpiece should not be placed on the edge of the worktable as it bears the risk of falling down.

iv. The workpiece should be positioned in such a way that the cut to be made is close to the vice. This ensures that the workpiece does not eject out from the vice which may cause injury.

v. The application of force while cutting and filing should be during the forward stroke only, whereas during the return stroke the pressure should be released.

vi. The workpiece should not be held in hand while working on it.

vii. The file should be held with the help of a properly fitted tight handle. Files without handle should never be used as it is unsafe for hands.

viii. The burrs formed on the edges of a workpiece should be removed by filing to prevent injury to the hand.

ix. The bench vice should not be used for hammering the workpiece.

x. The hacksaw blade should be kept straight while sawing, as it may break otherwise.

Table 5.1 Best practices
(RIGHT AND WRONG WAY)

Illustration of right and wrong way	Benefits of doing right
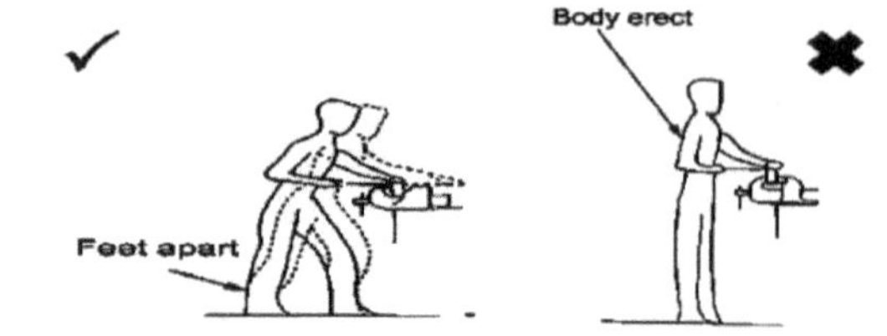	High pressure on the workpiece and low strain to hands and legs.
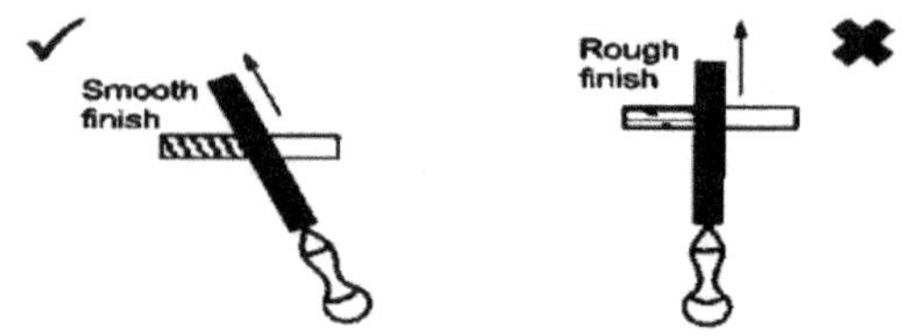	Filing diagonally across the workpiece results in smooth finish without scratch marks.
	The workpiece should be kept low in the vice to ensure it gets high rigidity and low vibrations.
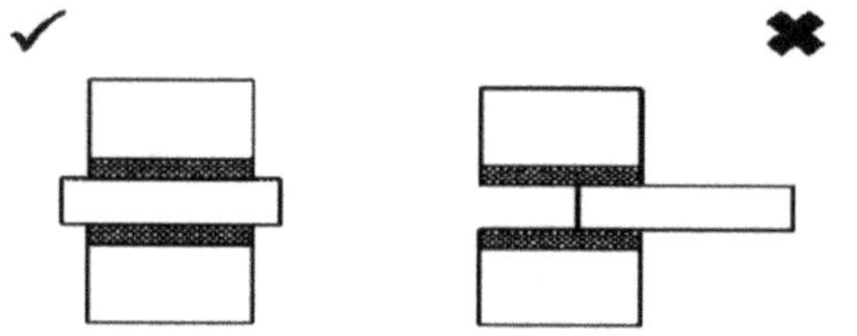	Hold the work within the width of the vice jaws to get better surface finish.
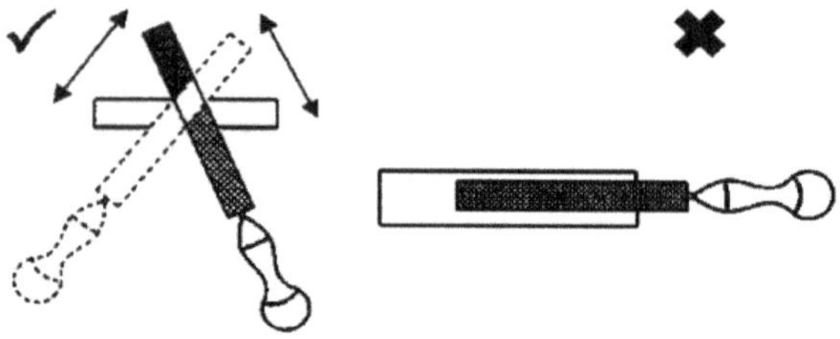	Work across at right or left angle. It is wrong to file along the length of the workpiece.

Right use of hacksaw	Remarks
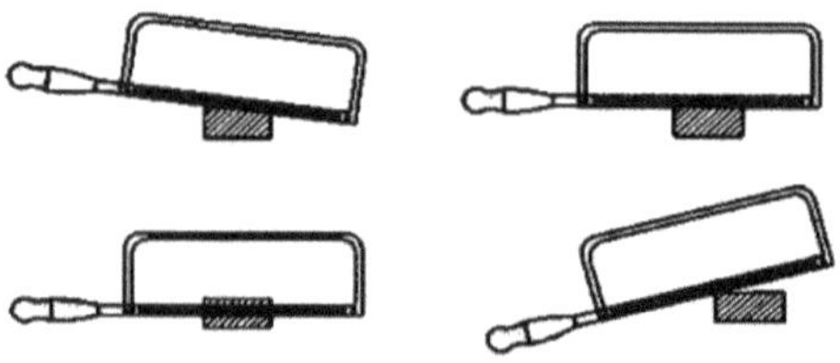	The hacksaw blade should be kept at slight inclination to start cutting. Later it can be brought to the horizontal position.

PRACTICE NO. 5.1

Job: Prepare a square workpiece of size 48 mm x 48 mm

Objectives: Learn by practice various fitting operations, such as filing, marking, punching, sawing and finishing to prepare a square plate workpiece.

Machines, equipment, and tools required: Bench vice, set of files, steel rule, try-square, vernier caliper, ball-peen hammer, scriber, dot punch, surface plate, angle plate and anvil.

Material required: Mild steel square workpiece of size 50 mm x 50 mm x 5 mm

Schematic of the job:

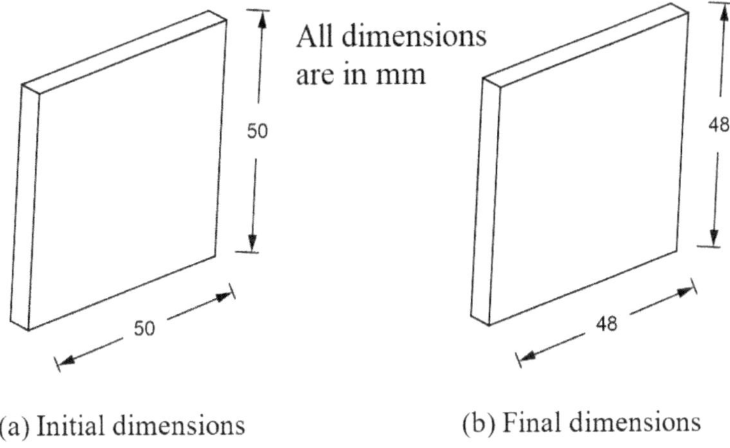

Figure P5.1: Schematic of the square shape job

Procedure:

i. The dimensions of the given workpiece are checked with a steel ruler.

ii. With the help of a hacksaw, cut a square shaped workpiece of about 50 mm x 50 mm x 5 mm size from a mild steel strip.

iii. The workpiece is fixed rigidly in a bench vice and its one side face filed using a rough flat file first and then a smooth flat file. Now file its side face, which is at a right angle to the side face filed first. Check the right angle between the two side faces with the help of a try square one by one.

iv. Similarly, file all side faces of the workpiece. File the workpiece in such a way that the right angle between the adjacent side faces is maintained. The right angle of the adjacent side faces is checked using a try square.

v. Measure length of the side faces of the job and also check the right angles with the help of a steel foot ruler and a try square multiple times as and when needed.

vi. Fill out the workshop practice response sheet given at the end of chapter 2 and answer the questions provided therein. Get feedback of the instructor on the workpiece prepared by you.

PRACTICE NO. 5.2

Job: To prepare a L shaped workpiece from the given mild steel piece

Machines, equipment, and tools required: Hand hacksaw, try square, chisel, dot punch, hammer, surface, bench wise, plate, steel foot rule, outside caliper, odd leg caliper, scriber, chalk, flat rough file, flat smooth file.

Material required: Two mild steel square workpieces of size 48 mm x 48 mm x 5 mm

Objectives: Learn by practice fitting shop operations, such as filing, marking, punching, sawing, and finishing to make a L shaped workpiece.

Schematic of the job:

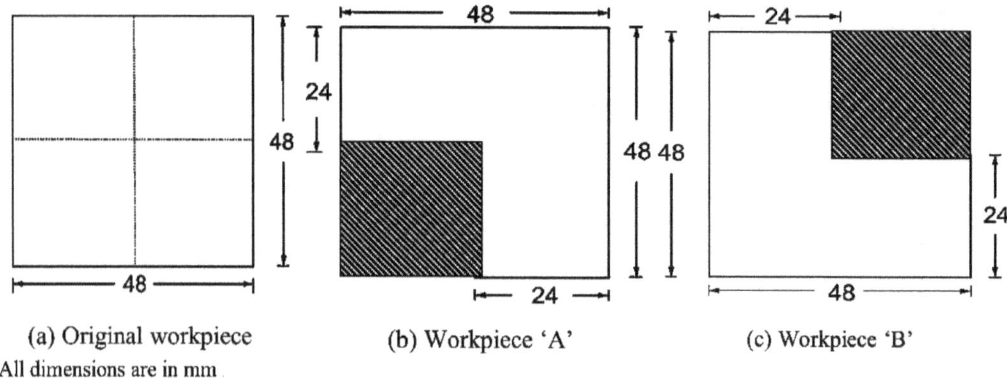

Figure P5.2 Schematic of the L shape job for fitting

Procedure:

i. Two mild steel workpieces as obtained from the practice 5.1 are checked with the help of a measuring scale.

ii. One of the workpieces is marked as 'A' is applied with chalk on its surface.

iii. With the help of a scriber two lines are marked at right angles to identify L shape portion shown as hatched area in Figure P5.2 (b). Measuring scale and try-square are used to help mark the lines.

iv. Dots are marked along the scribed lines using a dot punch.

v. Cutting action is performed along the scribed lines and punched dots to remove the required material shown with the help of hatching in the Figure P5.2 (b).

vi. All side faces of the workpiece are filed first with a rough flat file and then with a smooth file.

vii. The flatness of all the side faces is checked with the help of a precision try-square.

viii. Angle of the new cut side faces is also checked to verify that the cut sides are at right angles.

ix. Similar operations are done for the workpiece marked 'B' as shown in Figure P5.2 (c).

x. Both the workpieces (marked 'A' and 'B') are fitted together and held in a vice to file and finish on the faces of the workpiece.

xi. Fill out the workshop practice response sheet given at the end of chapter 2 and answer the questions provided therein. Get feedback of the instructor on the workpiece prepared by you.

PRACTICE NO. 5.3

Job: To prepare a square fit for the given two mild steel pieces

Objectives: Learn by practice various fitting shop operations, such as filing, marking, punching, sawing, and finishing to make two jobs with square fit.

Machines, equipment, and tools required: Hand hacksaw, try square, chisel, dot punch, hammer, surface, bench wise, plate, steel foot rule, outside caliper, odd leg caliper, scriber, chalk, flat rough file, flat smooth file.

Material required: Two mild steel square workpieces of size 45 mm x 45 mm x 5 mm.

Schematic of the job:

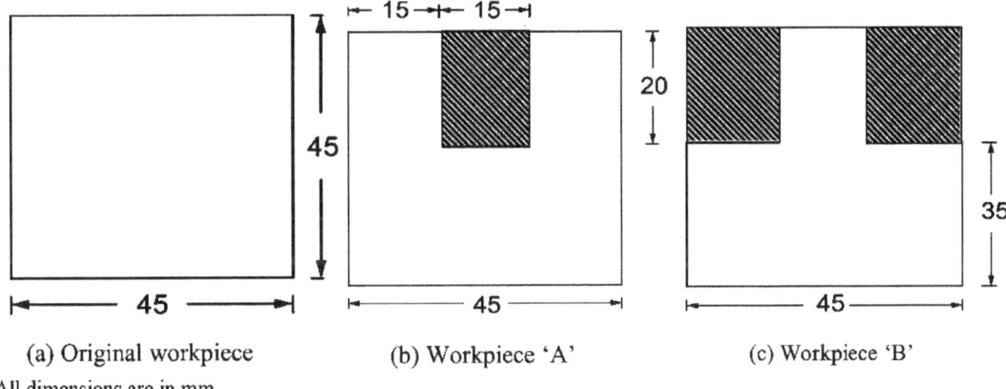

Figure P5.3 Illustration of a square fit job

Procedure:

i. The given mild steel workpieces are measured and made to size.

ii. The side with 45 mm length is filed using a rough flat file first followed by filing with smooth pipe.

iii. A precision try square is used to check the flatness of all the side faces and also check that they are at right angles.

iv. With the help of an odd leg caliper, markings are made on the workpieces 'A' and 'B', along the lines as shown in Figure P5.3.

v. Hacksaw is used to remove extra material from the workpiece 'A' shown as the hatched portion in the Figure P5.3. The markings on the workpiece are made with the help of a marking punch and odd leg caliper.

vi. Similarly, the excess material is removed from the workpiece 'B' with the help of a hacksaw.

vii. Filing is done on both the workpieces marked 'A' and 'B' to finish their mating side faces, removing burrs and making them right angled.

viii. Now both the workpieces marked as 'A' and 'B' are fitted together and held in the vice. Filing and finishing is further done if required on all the mating faces.

ix. Fill in the workshop practice response sheet given at the end of chapter 2 and answer the questions provided therein. Get feedback from the instructor on the workpiece prepared by you.

PRACTICE NO. 5.4

Job: To drill a hole and make threads by tapping on a mild steel piece.

Objectives: Learn by practice the drilling and tapping operations performed in the fitting shop.

Machines, equipment, and tools required: Hacksaw, try square, steel foot rule, dot punch, ball peen hammer, divider and twisted drill, tap, wrench, surface plate and centre punch.

Material required: Mild steel square workpiece of size 50 mm x 50 mm x 5 mm. Alternatively, job from any of the three previous practice sessions can be taken.

Schematic of the job:

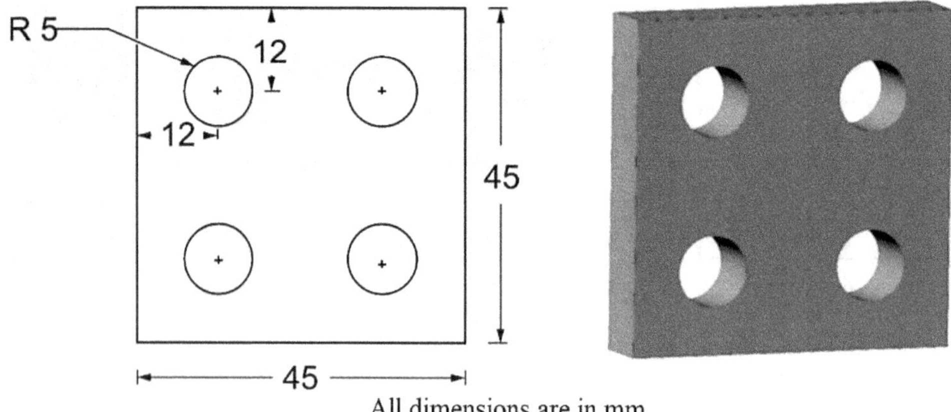

Figure P5.4 Schematic of the job for drilling and tapping of threads

Procedure:

i. Hold the mild steel strip in a bench vice and cut it to size as per the job description given in Figure P5.4. Alternatively, any of the jobs made in the previous three sessions may be used.

ii. Check squareness of the job with the help of a try square and measuring scale so that its side faces are at right angles.

iii. Place the job on the surface plate and mark four lines at 12 mm distance from each edge as shown in figure P5.4.

iv. Mark four centers with the help of a center punch at the intersection of four marked lines.

v. Take a twist drill of 10 mm diameter (5 mm radius) and fix it into the collet or drill chuck attached to the spindle of the bench drill.

vi. Fix the job in the machine vice or table of the bench drill.

vii. Produce four number of round holes in the workpiece with the help of marked centers as per the dimensions given in Figure P5.4.

viii. Take a tap and hold it in the tap wrench for producing internal threads in the drilled holes.

ix. Fix the workpiece in the bench vice, make internal threads in the holes by using a tap and tap wrench.

x. Clean the job with a wire brush.

xi. Fill out the workshop practice response sheet given at the end of chapter 2 and answer the questions provided therein. Get feedback of the instructor on the workpiece prepared by you.

Chapter 6

ELECTRICAL SHOP

6.1 INTRODUCTION

Electricity is an essential need in our daily life. It is supplied for domestic purposes in a single phase with a voltage of either 110 V or 230 V using a two-wire system. In an industrial setup, the electrical supply is instead delivered with a four-wire system to give 440 V in three phases. Figure 6.1 illustrates electrical wiring for an industrial as well as a domestic setup showcasing electricity supply using electric conductors and components. A wiring system is expected to facilitate specified flow of current and constant voltage.

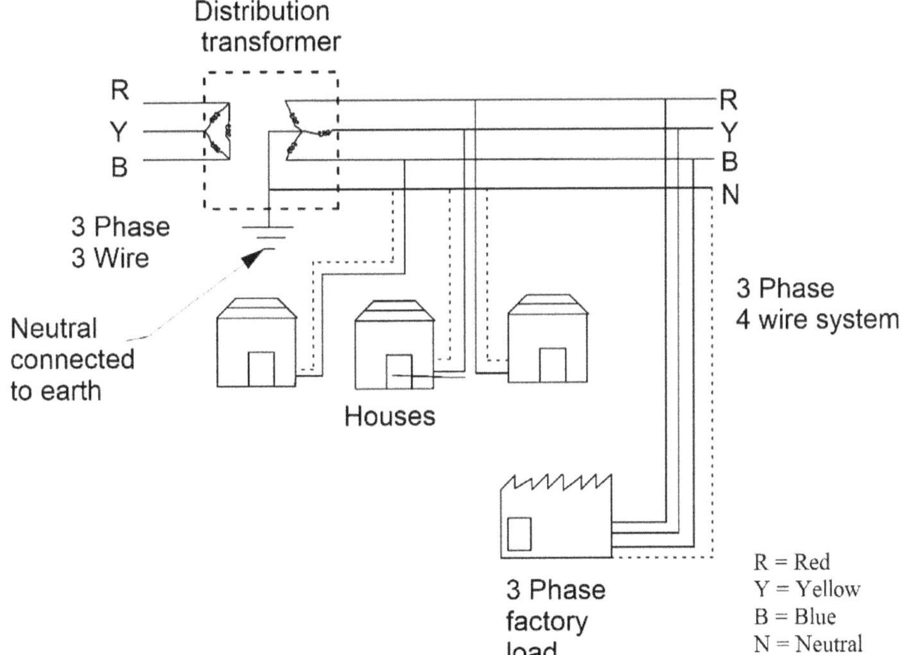

Figure 6.1 An illustration of three phase – four wire and single phase - two wire electrical supply

6.2 WIRES AND WIRE SIZES

An electric wire is made of a conducting material, which is covered with an insulation. The insulation material is generally made of vulcanized rubber or polyvinyl chloride. The electric wire may consist of one or several twisted strands, which are classified into three types, single core, double core and multi core.

Normally, the size of an electric wire is specified in terms of its diameter. An electrical wire with larger diameter has a higher current carrying capacity. Figure 6.2 (a) shows snapshot of a standard wire gauge (SWG). The wire gauge helps measure the gauge of the wire, which can be used to find out diameter of the wire using a conversion table, a snapshot of which is given in Figure 6.2 (b).

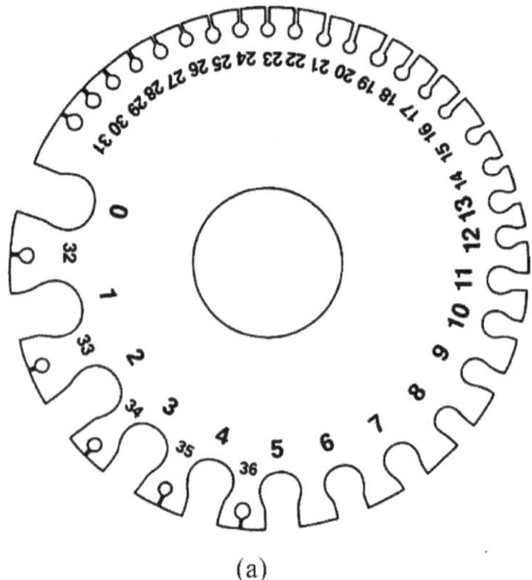

(a)		(b)

SWG	Diameter	
	Inch	mm
7/0	0.500	12.700
----	---	----
1/0	0.324	8.236
1	0.300	7.620
2	0.276	7.010
4	0.232	5.893
--	--	--
21	0.032	0.813
22	0.028	0.711
23	0.024	0.610
--	--	--
49	0.0012	0.030
50	0.001	0.025

Figure 6.2 (a) Standard wire gauge (SWG) (b) SWG number to diameter conversion table

6.3 CONDUIT PIPE AND ITS FITTING

Electrical conduits are pipes made of plastic, metal, or fiber for providing the required protection to a wiring system. It can be said that an electrical system cannot be laid out without the help of conduit pipes and hence they are found in every residential and commercial building. The conduit pipes are available in the market in different shapes and sizes. Further, different types of joints are also used for connecting conduit pieces.

6.4 ELEMENTS OF WIRING SYSTEM

Important elements of an electrical wiring systems are discussed in the following sections.

6.4.1 Fuses and Circuit Breakers

Fuses are used to provide protection to a circuit against excessive current. Now a days, electrical circuits make use of circuit breakers. In case something goes wrong with an electrical circuit or an equipment, the circuit gets short-circuited which causes blowing out of the fuse or tripping of the circuit breaker. This results in isolation of the circuit or equipment hence making it safe. However, in case of a short circuit there is a need to check the circuit or the appliance to identify any defects to take corrective action.

Nowadays miniature circuit breakers (MCBs) are available in the market. MCB is an electromechanical device that protects an electric circuit in case excessive current flows through it. MCBs are normally classified according to the tripping current and the operating time. The current rating of MCBs ranges from 5 to 60 ampere. Figure 6.3 shows snapshots of a fuse, circuit breaker and an MCB.

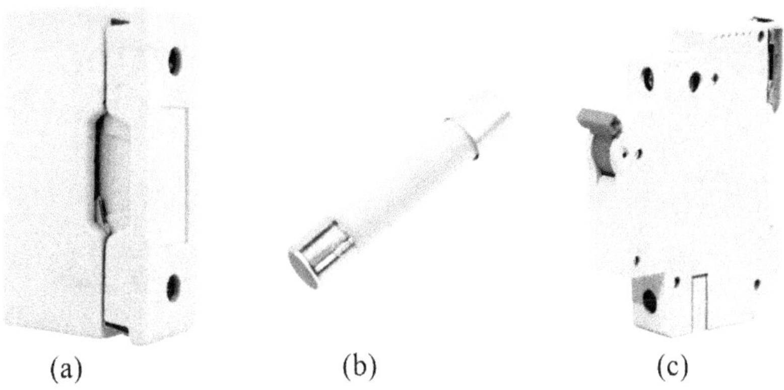

(a)   (b)   (c)

Figure 6.3 Snapshots of (a) fuse, (b) circuit breaker, and (c) MCB

6.4.2 Switches

A switch is used for starting the supply of electricity to an electrical equipment and also to stop the supply when not needed. Therefore, switches are binary devices which have two states, i.e., on and off. Different types of switches are available in the market, the important types are below mentioned.

i. *One way switch or single-pole electrical switch* (Figure 6.4 (a)) is a very commonly used switch, which has ON and OFF markings. This is used for control of fans, lights or other electrical devices.

ii. *Power switches* (Figure 6.4 (b)) are used when a higher current is required for the electrical device. For example, air conditioners and geysers generally require power switches with a current rating of 15 Ampere, which cannot be safely provided with normal 5 Ampere switches.

iii. *Two-way switch* (Figure 6.4 (c)) is used when an electrical device is to be controlled from more than one location. It facilitates the users to switch off or on the device from either location.

iv. *Do not disturb switch* (Figure 6.4 (d)) is used to provide an indication to the persons to help ensure privacy and silence.

v. *Light dimmer switch* (Figure 6.4 (e)) is used to control the intensity of current being supplied to the electrical device. The main application of a light dimmer switch is to change the light intensity of a bulb or to control the speed of a fan.

vi. *Bell push switch* (Figure 6.4 (f)) is used for ringing the bell in homes, buildings and offices. The concept used in bell push switch is simple, to bring the circuit to the 'ON' position when pushed and thereafter immediately, with the help of a spring mechanism it comes to the 'OFF' position.

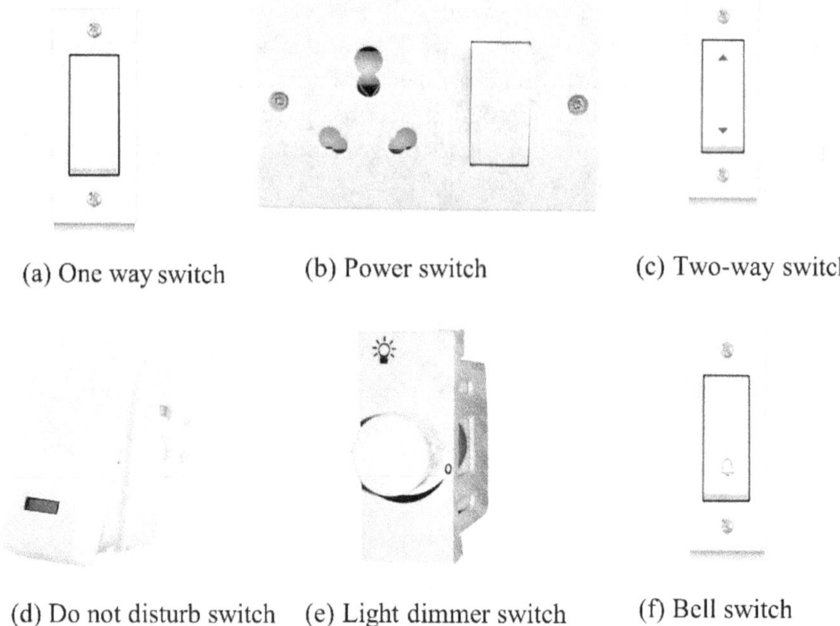

(a) One way switch   (b) Power switch   (c) Two-way switch

(d) Do not disturb switch   (e) Light dimmer switch   (f) Bell switch

Figure 6.4 Electric switches

6.4.3 Distribution board

A distribution board provides the required space for the assembly of fuses and circuit breakers and is used to connect the incoming power supply with the electrical circuits laid out in a building. The main purpose of the distribution board is to provide control of the electrical circuits at one place for safety. Figure 6.5 shows a snapshot of a distribution board.

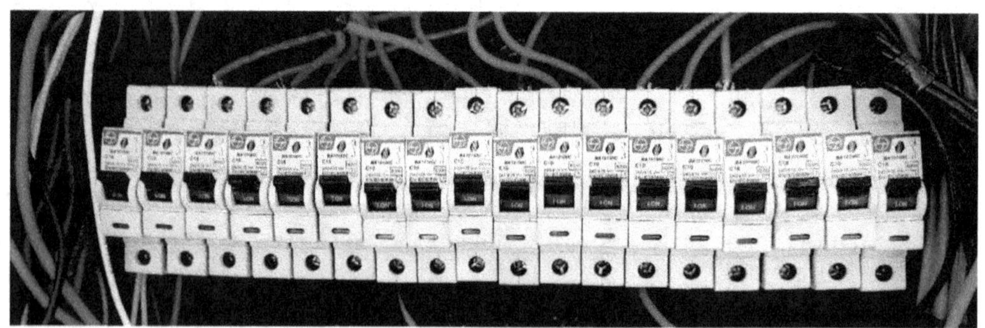

Figure 6.5 Distribution board

6.5 SYMBOLS USED IN ELECTRICAL CUIRCUITS

Drawing of electrical circuits requires using different symbols. Figure 6.6 provides important electrical symbols used in drawing of electrical circuits.

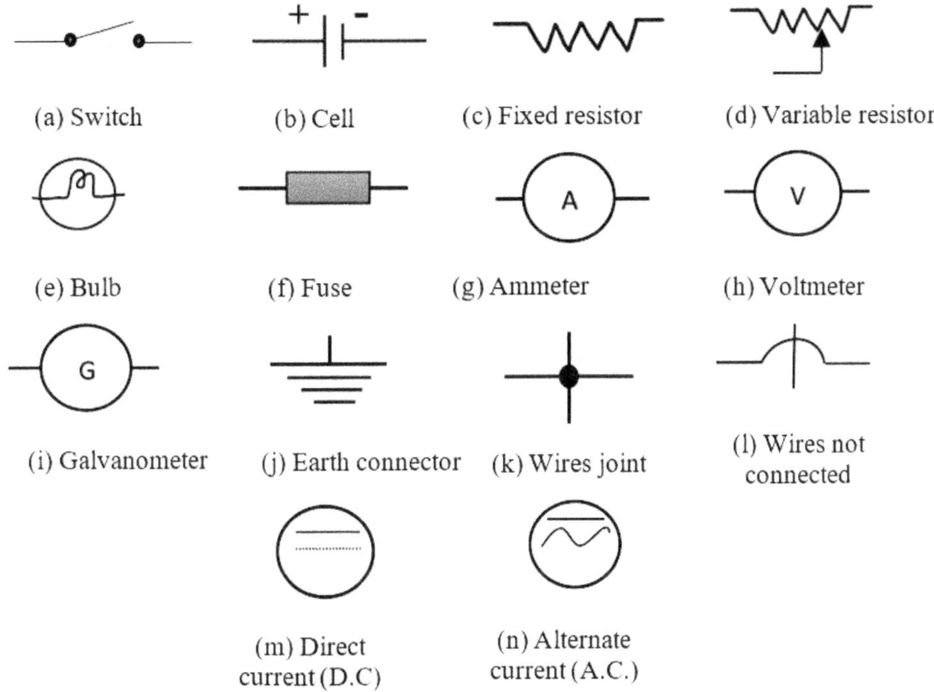

Figure 6.6 Symbols used in electrical circuits

6.6 WIRING METHODS

Mainly two types of electric circuits are used in domestic scenarios, namely series circuit and parallel circuit. In a series circuit, two or more electrical devices or components share a common node. Therefore, in a series circuits, same current flows through each of the electrical devices. On the other hand, in parallel circuit, the electrical current is distributed amongst multiple paths. Therefore, different current flows in each device, which means that in parallelly connected electrical devices, the current is divided, however, the voltage supplied to each device is the same. Figure 6.7 shows a snapshot of series and parallel circuit.

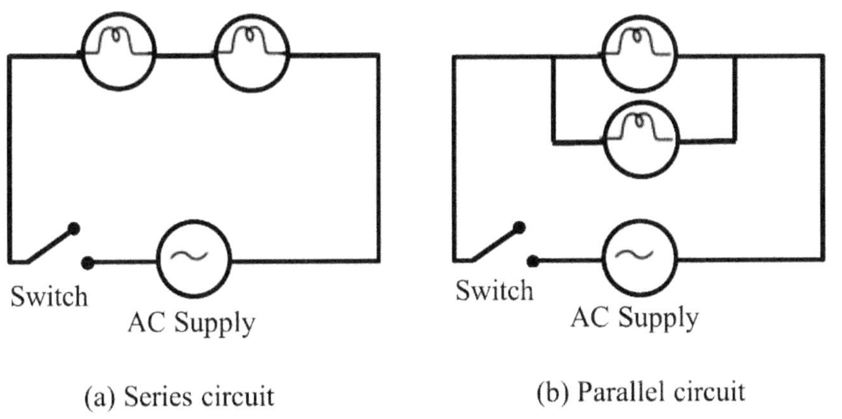

(a) Series circuit   (b) Parallel circuit

Figure 6.7 Series and parallel circuit wiring methods

6.7 TOOLS AND MATERIALS

An electrical shop has several tools and material which can be used for preparing a wiring circuit. Figure 6.8 shows snapshots of these tools and material along with a brief explanation of their use.

Screwdriver with an insulated handle is used for tightening or loosening of screws and nuts by turning the screw head.

Wire stripper is used to uncover the insulation covering from the electrical cables, which is required for making electrical connections.

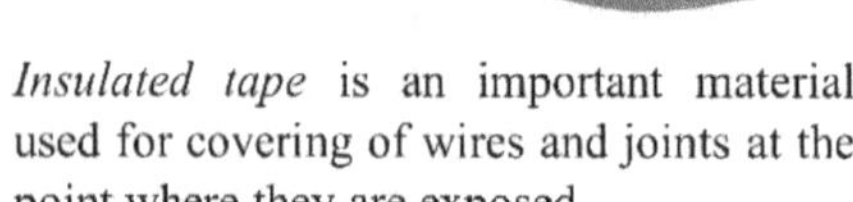

Insulated tape is an important material used for covering of wires and joints at the point where they are exposed.

Plier is used for many tasks in electrical shop that require application of torque which otherwise is not available with bare hands.

Spanner is required to loosen or tighten nuts, bolts, or similar fixings.

Hacksaw is used for cutting of metallic pieces or conduits.

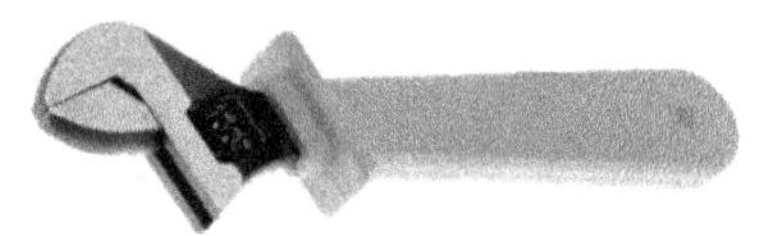

Voltage tester is used to check presence of voltage in a wire or conducting surface.

Soldering iron is used for heating of soldering alloy for making soldering joints.

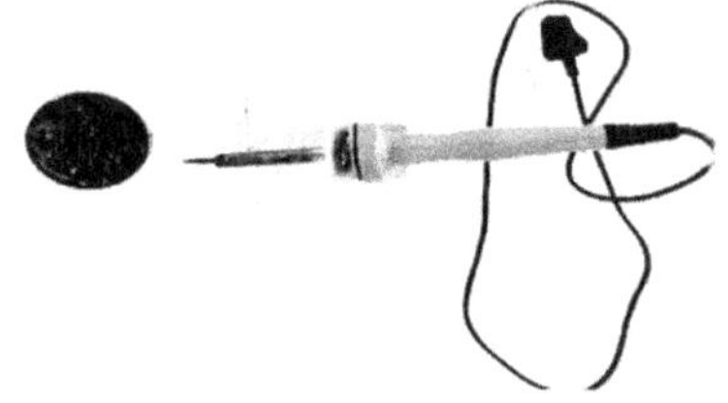

Figure 6.8 Tools and materials for electrical circuits

6.8 SAFETY PRECAUTIONS

Following safety precautions in electrical shop are required to be followed.

i.   While operating an electrical switch, grasp it from the insulated handle only.

ii. Avoid running multiple electrical devices from one point.

iii.   Make use of circuit breakers or fuses of proper capacity to protect the devices and avoid danger.

iv.   During repair and maintenance of electrical circuits and devices, disconnect it from power supply. Never work on electric wires when the power is on.

v.   In case of fire from electrical circuits, don't use water as it is electrically conducting, instead use sand.

vi.   In case of power failure, keep all devices and equipment switched off to prevent sudden power demand and fluctuation, in case the power supply resumes.

vii.   While removing plug from the socket, don't hold it from the wire, hold the plug instead.

viii.   Never work on electrical circuit with bare feet, always wear electrically insulating shoes.

ix.   Use one hand only to hold the tester, whereas the second hand should be kept in the pocket. This will avoid electrical shock as the chances of second hand accidently touching earth, wall or another conducting material is minimized.

x.   An electrical device should be properly checked before providing electrical power to it.

xi.   Earth connection of an electrical device should be checked before switching-it on.

xii.   Before replacing the blown fuse, always switch-off the main switch.

PRACTICE NO. 6.1

Job: Make electrical connections to connect two lamps in a series electrical circuit.

Objectives: Learn by practice electrical wiring and connections for making a series electrical circuit.

Machines, equipment, and tools required: Tester, screwdriver, AC power supply, plier, hammer.

Material required: Wooden wiring board, one way switch, round wooden block, batten, lamp holders, connector, screwdriver, wires, wire clips, nails, wood screws and lamps (or bulbs), insulated tape.

Schematic of the job:

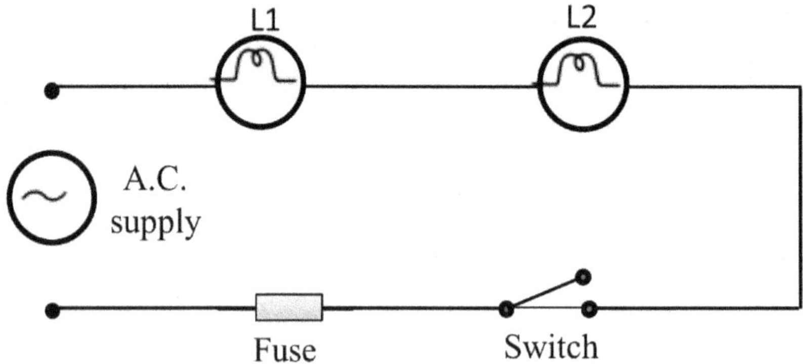

Figure P6.1 Schematic of a series electrical circuit

Procedure:

i. Mark wiring diagram on the wooden board.
ii. Nail the clips into the wooden board according to the wiring diagram.
iii. Stretch the wires and clamp them with the help of clips.
iv. Screw round blocks, three in number, onto the board as per the wiring diagram.
v. Connect the wires to the holders and switches.
vi. Fit the bulbs to the holders.
vii. Test the wiring connection by giving power supply in supervision of the instructor.
viii. Fill in the workshop practice response sheet given at the end of Chapter 2 and answer the questions provided therein. Take feedback of the instructor on the job you have made.

PRACTICE NO. 6.2

Job: Make electrical connections to control a lamp with the help of a two-way switch.

Objectives: Learn by practice the procedure of making a two-way switch.

Machines, equipment, and tools required: Tester, screwdriver, AC power supply, plier, hammer.

Material required: Wooden wiring board, two-way switches, wooden round block, batten lamp holder connectors, screwdriver, wires, wire clips, nails, wood screws and bulbs, insulating tape, hammer.

Schematic of the job:

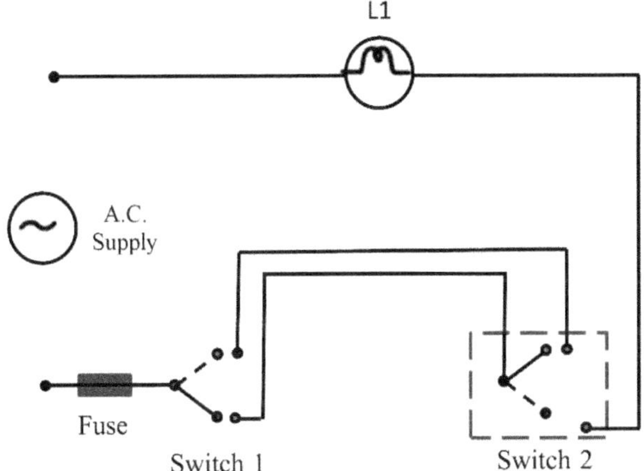

Figure P6.2 Circuit diagram for a two-way switch

Procedure:

   i.   Mark wiring diagram on the wooden board.

   ii.   Nail the clips into the wooden board according to the wiring diagram given.

   iii.   Stretch the wires and clamp them with the help of clips.

   iv.   Screw round blocks, three in numbers, onto the board as per the wiring diagram.

   v.   Connect the wires to the holder and two-way switches and screw them into the round blocks.

   vi.   Fit the bulb to the holder.

   vii.   Test the wiring connection by giving power supply under supervision of the instructor.

   viii.   Fill in the workshop practice response sheet given at the end of Chapter 2 and answer the questions provided therein. Take feedback of the instructor on the job you have made.

PRACTICE NO. 6.3

Job: Make electrical connections to assemble a fluorescent lamp with its accessories.

Objectives: Learn by practice the assembly of different parts of a fluorescent lamp and connect them as per the required circuit.

Machines, equipment, and tools required: Tester, screwdriver, AC power supply, plier, hammer.

Material required: fluorescent tube, tube base, starter, choke, wire and screwdriver.

Schematic of the job:

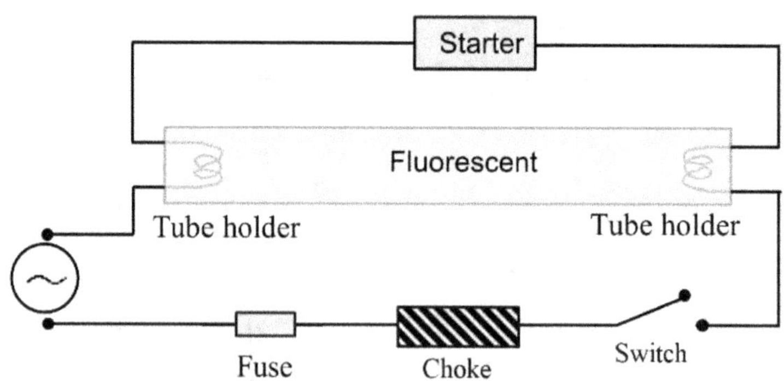

Figure P6.3 Circuit diagram of a fluorescent tube

Procedure

i.	Mark the wiring diagram on the table.
ii.	Fix the tube holder and the choke on the tube base.
iii.	Fix the fluorescent tube with the holders.
iv.	Finally connect the starter in series with the tube as per the given circuit diagram.
v.	Test the wiring connections by giving power supply.
vi.	Fill in the workshop practice response sheet given at the end of Chapter 2 and answer the questions provided therein. Take feedback of the instructor on the job you have made.

Chapter 7

ELECTRONICS SHOP

7.1 INTRODUCTION

Electronics shop is used for the fabrication of products that require electronic components and circuits. The following sections discuss important components used in the electronic shop.

7.2 ELECTRONICS COMPONENTS

In this section important electronic components, namely resistors, capacitors and inductors are discussed.

7.2.1 Resistors

Resistors (Figure 7.1 (a)), very commonly used in electronic circuits are passive components that have two terminals. The major function of a resistor is to provide electrical resistance in a circuit. Resistors are color coded, however, some of the resistors have their resistance values printed in Ohms. Multi-meter, which is a commonly used device for electronics can be used to check the resistance of a resistor. On a circuit board, a resistor is marked with the letter "R".

7.2.2 Potentiometer

A potentiometer (Figure 7.2 (b)) is a type of resistor in which resistance can be varied. The potentiometers available in the market have marking showing the maximum value of the resistance in Ohms. Potentiometers may use a three-digit code in which the last digit is used as a multiplier. For example, code 204 equals 20 followed by four 0's, which means 200000 Ohms.

Furthermore, potentiometer also has a letter code indicating the variable rate of resistance, which is marked as 'VR' on a circuit board.

7.2.3 Capacitors

A capacitor also known as the condenser is a passive electrical component which stores energy electrostatically. Capacitors are very commonly used electronic components and have two terminals. Specifications of a capacitor are indicated by its capacitance value printed on it with the help of a three-digit code, however some capacitors are color coded. Capacitors are generally marked with the letter "C" on a circuit board. Figure 7.1 (c) shows a snapshot of capacitors.

7.2.4 Inductors

An inductor (Figure 7.1 (d)) is a passive component, which is also known as a coil, or a reactor. It comprises a conductor such as a wire, which is wound in the form of a coil. The function of an inductor is to provide resistance when current changes. The main principle on which an inductor functions is that with current flowing through it, the energy is stored in a magnetic field in its coil. Inductors may be color coded to mention their inductance rating. A device which can measure inductance is also helpful to find the inductance. In the circuit diagram, the inductors are marked with the letter 'L'.

(a) Resistors   (b) Potentiometer   (c) Capacitors

(d) Inductor   (e) Transformer   (f) Fuse

(g) Diode   (h) Transistor   (i) Bridge rectifier

(j) Integrated circuit   (k) LED indicators   (l) Switch

(m) Cell   (n) Relays

Figure 7.1 Electronic components

7.2.5 Transformer

A transformer (Figure 7.1 (e)) is a device which is very commonly used in electrical and electronic devices. However, the transformer used in electronic devices is meant to transmit less power in comparison to those used in electrical circuits. It makes use of inductive coupling between its winding circuits to transmit electrical power and alter the voltage as per the requirement. The transformers which help in reducing the voltage are known as step down, whereas others that increase the voltage are called step up transformers. Transformers are easily identifiable and have their specifications mentioned. The letter "T" is marked on a circuit board to represent a transformer.

7.2.6 Fuses

Fuses are used in both electronic and electrical circuits as a sacrificial device to protect against overcurrent. An essential component in a fuse is a strip or a metal wire which is melted when more than a threshold current passes through it. This helps in breaking the circuit and prevents flow of current to save the equipment. The primary reason for excessive current is overloading, mismatched loads or device failure. Figure 7.1 (f) shows a snapshot of fuses.

7.2.7 Diodes

A diode (Figure 7.1(g)) is an electronic component which has two terminals with asymmetric conductance. The resistance of a diode in one direction is very low, almost amounting to zero, whereas in the other direction the resistance is very high. The letter "D" is used to represent a diode on a circuit board.

7.2.8 Transistors

A transistor (Figure 7.1 (h)) is a semiconductor device which is used for the amplification and switching of an electronic signal and electrical power. It has three terminals which are attached to the semiconductor material. The three legs of the transistor are used to connect it to the external circuit. The letter "Q" is used to mark a transistor on a circuit board.

7.2.9 Bridge rectifiers

A bridge rectifier (Figure 7.1 (i)) consists of four or more diodes arranged in a bridge circuit configuration, for converting alternating current (AC) to the direct current (DC). On a circuit diagram, the letters "BR" is used to notate the Bridge Rectifiers.

7.2.10 Integrated circuit (IC)

An integrated circuit also called IC in the common language (or simply chip or micro-chip) is composed of a set of electronic circuits on one small plate (also called chip) of semiconductor material, which is normally silicon. An IC can be significantly reduced in size when compared with a discrete circuit made from independent components. ICs are nowadays commonly used in virtually all electronic equipment. The letter "U" or "IC" is used to represent integrated circuits on a circuit diagram. Figure 7.1 (j) shows a snapshot of an integrated circuit.

7.2.11 LED and LED display

A light-emitting diode, which is also called LED, is also a semiconductor which can emit light. LEDs are used for providing indication in a device. Generally, LEDs emit low-intensity color light, however, modern LEDs are capable of emitting visible, ultraviolet, and infrared wavelengths with high brightness. LED displays are a panel of LEDs which can be used as display screens. Figure 7.1 (k) shows a snapshot of LED.

7.2.12 Switches

Switch (Figure 7.1 (l) is a component or device which can be used for breaking or making of an electrical circuit.

7.2.13 Batteries

A battery (Figure 7.1 (m)) is a device which can store electrical energy and release it when required. It has one or more electrochemical cells which are used to convert chemical energy into electrical energy. Batteries are marked with specifications, which can very easily be identified.

7.2.14 Relays

A relay is a kind of a switch which is operated electrically. Generally, relays make use of an electromagnet for mechanically operating a switching mechanism, however, it may also use other operating principles. Relays are used when a circuit is to be controlled by a low-power signal or several circuits are required to be controlled by one signal. Relays have their specifications mentioned on them. The letter "K" is marked on a circuit board to represent a relay. Figure 7.1 (m) shows a snapshot of relays.

7.3 TESTING OF ELECTRONICS COMPONENTS

A Multimeter is an instrument which is the most sought after for working on electronic circuits. A multimeter can be used for measuring several important parameters, such as voltage, current and resistance, which is also the reason that it is called a multimeter. Some models of multimeter also provide additional functions, such as transistor test, diode test, continuity test, transistor test, and frequency test. Figure 7.2 shows a snapshot of multimeter.

A multimeter has three parts, namely display, selection knob and ports. The display panel of the multimeter generally displays the measured parameters in digital or analog format with a provision to indicate negative values. The selection knob has a provision to set the multimeter to show different parameters, such as milliamps (mA) of current, voltage (V) and resistance (Ω). A multimeter has two probes which can be plugged into two ports on the front unit. The important tasks that can be done with the help of a multimeter are as follows.

- Voltage (AC or DC) measurement
- Measurement of current
- Resistance measurement
- Current measurement

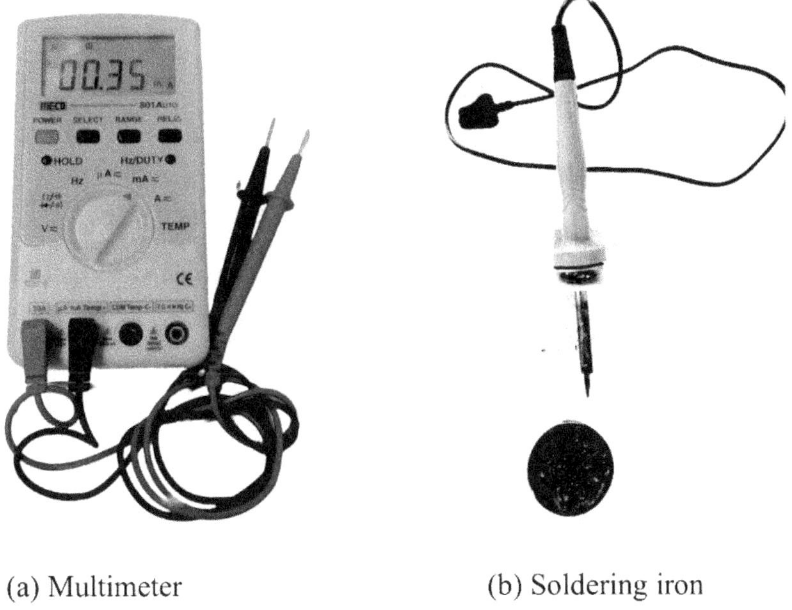

(a) Multimeter   (b) Soldering iron

Figure 7.2 Snapshot of a multimeter and soldering iron

The readers are encouraged to identify the components and tools discussed in this chapter until now and learn the use of multimeter to measure parameters for electronic circuits.

7.4 SOLDERING

Soldering is one of the important tasks frequently used in an electronic shop for making joints. This involves joining of electronic components in circuits making use of fusion of relatively low melting point alloys, which is called Solder or Soldering alloy. The Soldering alloy is made of Tin and Lead. The alloy has a low melting point, and it adheres to the surfaces, which are required to be soldered or joined. The joint made using soldering is a semi-permanent joint which can be broken using minimal force without damaging the electronic components. The following paragraphs discuss soldering equipment and procedure.

7.4.1 Soldering equipment and materials

There are two main items required for soldering, namely soldering iron, and solder. Soldering iron has a heating element and is the source of heating to melt the solder or soldering alloy. Soldering irons are available with a power rating of about 20 W that provide sufficient heat for normal soldering operations. Figure 7.2 (b) shows a typical soldering station.

7.4.2 Soldering Procedure

The following procedure is used for making a soldering joint.
 i. Clean the surface of the electronic circuit to be joined with the lead. This is very important for obtaining a strong soldering joint having low resistance.

ii. Bend the lead as per the requirement and place it on the point on the circuit board where it is to be joined.

iii. Plug the soldering iron keeping it on the docking station and allow it to be hot.

iv. An end of soldering wire is applied to the tip of the hot iron and the lead to be joined is placed alongside. This helps conduct the heat to the solder as well as end of the lead.

v. Once the lead and soldering wire get sufficiently heated, end of the soldering wire gets melted sticking to the hot soldering iron.

vi. The molten soldering alloy is applied to the end of the lead and the point where it needs to be joined on the circuit board.

vii. Allow the molten soldering alloy to solidify thus completing the job.

7.5 SAFETY PRECAUTIONS

i. The working environment should be well ventilated and illuminated.

ii. It is dangerous to leave soldering iron unattended, therefore switch it off when not in use.

iii. Clothes, hair, and power cables should be kept away from the soldering iron.

iv. The soldering iron should be returned to the soldering station very carefully ensuring that it is secure and safe.

v. The soldering iron should be held using the plastic handle.

PRACTICE NO. 7.1

Job: Identify various electronic components and make use of multimeter to find voltage, current, resistance of an electronic circuit.

Objective: Learn how to recognize different electronic components and find out different parameters of an electronic circuit using a multimeter.

Machines, equipment, and tools required: Electronic components, such as resistor, capacitor, transistor, diode, and transformer, Multimeter.

Schematic of the electronic circuit:

Figure P7.1 Electronic components soldered on a printed circuit board

Note: Fill out the electronics response sheet that follows and answer the questions provided therein. Get the feedback of the instructor on the work done by you.

ELECTRONICS SHOP RESPONSE SHEET

Student name: Roll number:

Job Name: Job material:

Objective of this practice:

List the electronic components identified by you in the electronic circuit board, state their functions and make their sketch.

Electronic component   *Function*   *Sketch*

Write down the parameters measured by you with the help of the multimeter.

Instructor remarks

Date: Signed/Checked by

PRACTICE NO. 7.2

Job: To connect a wire with the electronic circuit board using soldering

Objectives: Learn by practice soldering for making a joint of wire and electronic circuit.

Machines, equipment, tools, and material required: Soldering iron, soldering flux, soldering wire, wires, wire strippers and brass wool.

Schematic of the job:

Figure P7.2 Soldering

Procedure

i. Clean the soldering iron tip and tin it with the solder.

ii. Twist the ends of the wires together to combine them.

iii. Put rosin flux on the wire end so that it adheres to the solder properly.

iv. Heat the soldering wire end with the help of tip of the soldering iron.

v. Apply molten solder on the joint of the two electrical wires.

vi. Allow the solder to cool for about 1-2 minutes so that it solidifies.

vii. Clean the soldering iron and tin the tip. Try to keep the tip well tinned with a nice shiny layer of solder at all the time.

viii. After soldering, there can be excess of flux on the board and around the soldering joint, which needs to be cleaned with the help of a toothbrush and alcohol. This also helps to protect the joints from corrosion.

ix. Fill out the workshop practice response sheet given at the end of chapter 2 and answer the questions provided therein. Get the feedback of the instructor on the work done by you.

Chapter 8

SMITHY AND FORGING SHOP

8.1 INTRODUCTION

Smithy and forging shop is very useful for performing a number of operations, which mostly consist of heating the metal to bring it to the plastic stage followed by giving it the desired shape by forceful action such as hammering and pressing. Most of the operations carried out in this shop are done manually, therefore the part being produced are generally small and light. The hand forging process is generally employed for relatively small components.

8.2 TOOLS AND EQUIPMENT

Several tools and equipment are used in the smithy shop, which are explained in the following paragraphs.

8.2.1 Hearth

Hearth or forge is a primary equipment in the smithy shop required for heating the object to the desired temperature. The main structure of the hearth is made of cast iron or cast steel, whereas its legs are made of steel. The hearth has a chimney that allows the passage of flue gases. It also has an inlet opening for allowing air supply into the furnace. Fire bricks are used to make a lining in the areas exposed to fire. A blower is also used to supply the air for burning of fuel for heating. A water tank is also provided on the front so that the workpiece can be quenched after working for cooling. Figure 8.1 shows a snapshot of a smith's hearth.

Smithy or forging operation are carried out by bringing the metal to a temperature where it will become sufficiently soft as it comes to the plastic stage; the temperature however is lower than the melting point of the metal or alloy. Table 8.1 shows the temperature for carrying out forging or smithy operations for commonly used metals.

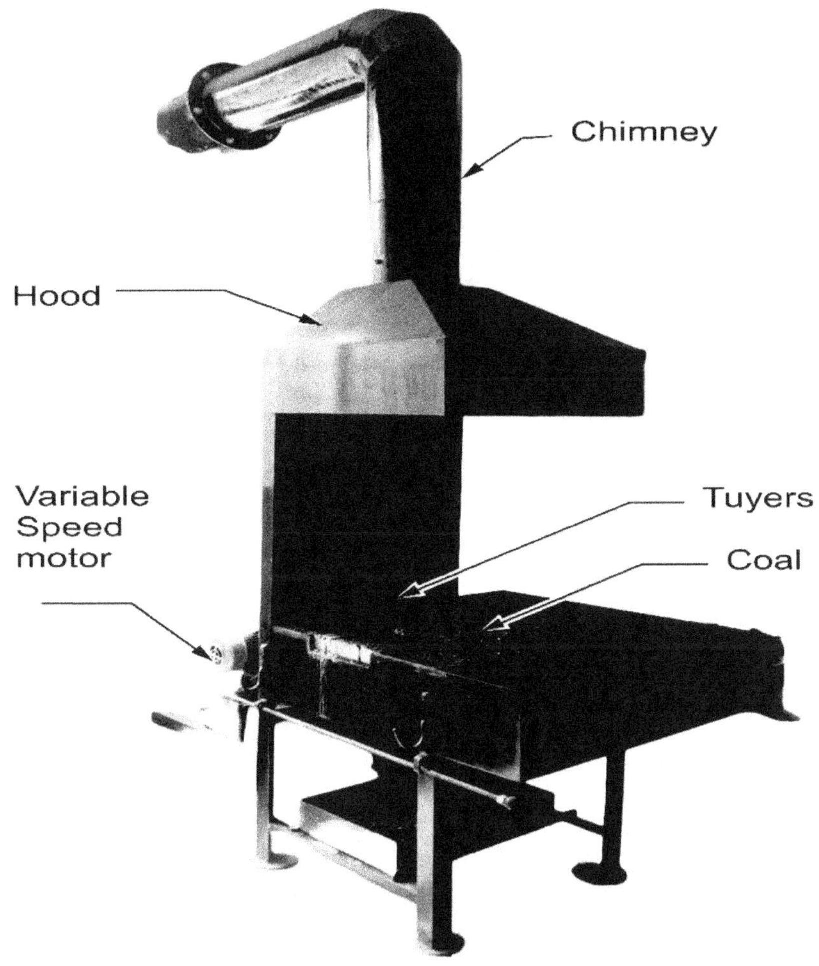

Figure 8.1 A snapshot of smith's hearth

Table 8.1 Forging temperature of metals

Metal	Forging temperature (°C)
Mild steel	760-1300
Wrought iron	910-1300
Medium carbon steel	750-1260
High carbon and alloy steel	810-1150

8.2.2 Anvil

Anvil is the most important tool in a smithy or forging shop. The main function of an anvil is to provide support to the job to work on it in the hot conditions. It has two holes named as *hardie* and *pritchel* on its top surface. The *hardy* hole has square shape for holding square shanks of swages and fullers. The *pritchel* hole is of round shape which can be used for bending of rods of small diameter and as a die for circular shapes and bending of rods of small diameters. Figure 8.2 shows a schematic of the anvil and its various parts.

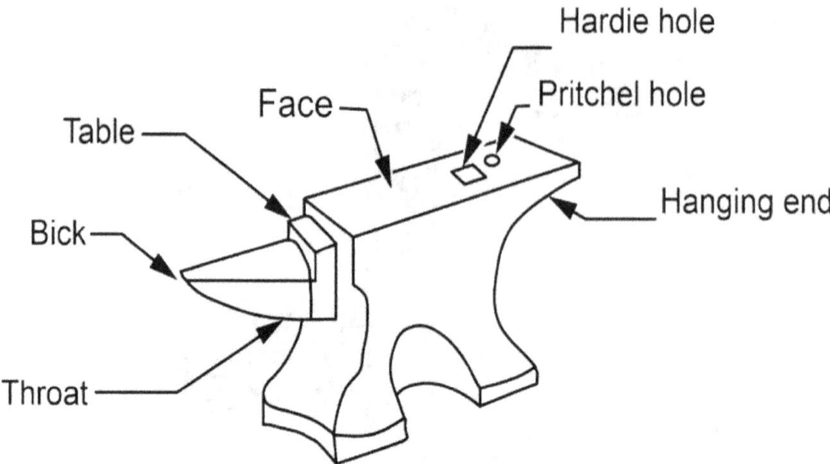

Figure 8.2 A schematic showing anvil and its parts

8.2.3 Swage block

Swage block is made of cast steel and has slots of different shapes and sizes on all its faces for facilitating smithy operations. It has a number of holes on its top face that go through its bottom face. The main operations done on a swage block are sizing, squaring, bending, punching, and forming. The swage block is specified in terms of block size and weight. Figure 8.3 shows a snapshot of a swage block.

Figure 8.3 A swage block

8.2.4 Hammers

The main striking tools in smithy shop are hammers, which are made of forged steel making them tough and strong. Hammers may be divided into two types, namely hand hammer and sledgehammer. Hand hammer is light in weight and therefore is used for light duty work, whereas the sledgehammer is heavy and therefore used for heavy duty work. Depending upon the shape, hand hammers are classified as ball peen hammer, cross peen hammer and straight peen hammer. Figure 8.4 shows snapshots of different types of hammers used in the foundry shop.

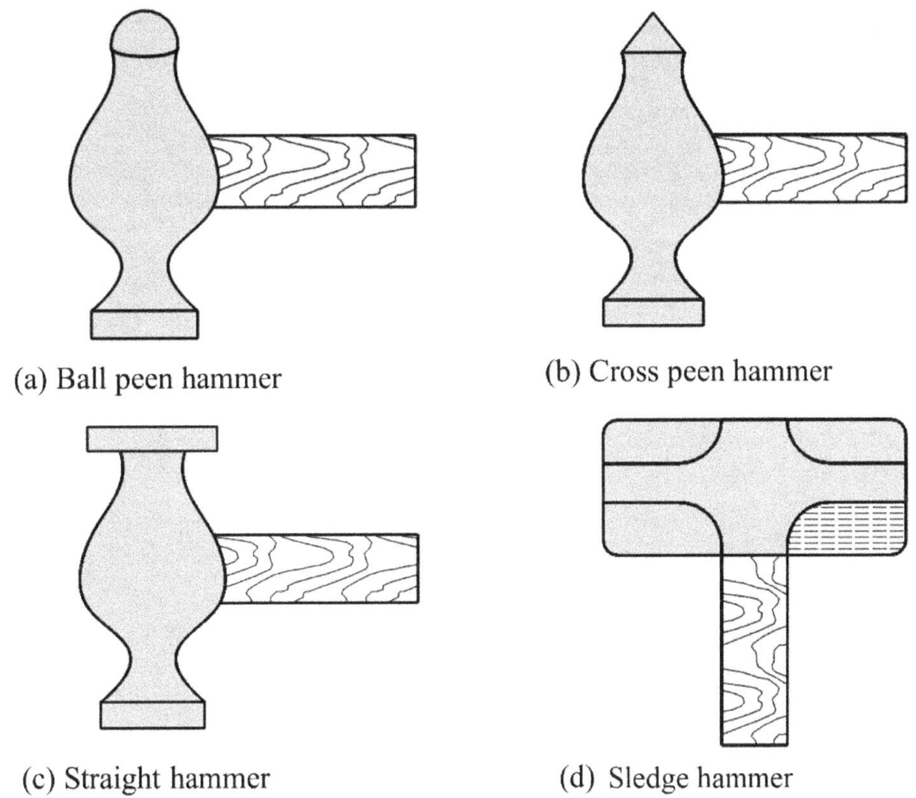

(a) Ball peen hammer   (b) Cross peen hammer

(c) Straight hammer   (d) Sledge hammer

Figure 8.4 Types of hammers used in smithy shop

Ball peen hammer has a round spherical end, which is commonly used for riveting and chipping. The cross-peen hammer has a pointed edge peen which is used for bending. The straight peen hammer has a flat surface peen which is used for stretching the workpiece. The sledgehammer is heavy and therefore used for heavy duty work.

8.2.5 Tongs

The workpieces in a smithy shop remain hot most of the time, which cannot be handled with bare hands. Therefore, for holding jobs during smithy operations, tongs, which are made of mild steel are commonly used. The following paragraphs discuss different types of tongs used in the smithy shop and their schematics are provided in Figure 8.5.

i. *Closed mouth tong* (Figure 8.5 (a)) is used for holding thin sections.
ii. *Open mouth tong* (Figure 8.5 (b)) is used for holding thick sections.
iii. *Round hollow tong* (Figure 8.5 (c)) is used for holding square, hexagonal, and orthogonal work.
iv. *Square hollow tong* (Figure 8.5 (d)) is used for holding round sections.
v. *Pick - up tong* (Figure 8.5 (e)) is used for picking of round bars. However, it is not used for holding workpieces during forging.

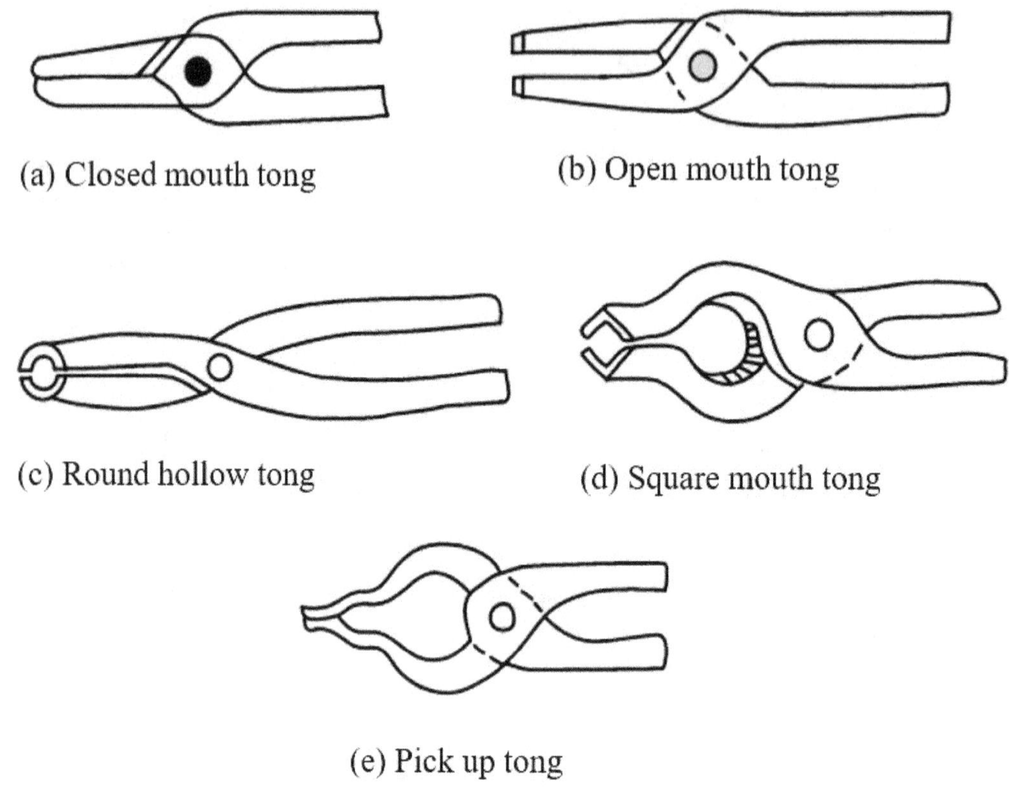

(a) Closed mouth tong   (b) Open mouth tong

(c) Round hollow tong   (d) Square mouth tong

(e) Pick up tong

Table 8.5 Types of tongs and their applications

8.2.6 Leg vice

Leg vice is generally used for holding workpieces for light forging and bending work. Like other kinds of vice, the leg vice is fixed to the workbench or a fixed structure.

8.2.7 Hardie

Hardie (Figure 8.6 (a)) is fitted in a hole provided at the top surface of the anvil, which is also called a hardie hole. The hardie has a cutting edge at its top, which is used for cutting and shearing operations along with the chisels. It is made of high carbon steel.

8.2.8 Fullers

Fuller (Figure 8.6 (b)) is used for forming of the hot metal workpieces. This tool generally comes in pairs, which is also called bottom and top fuller. Fuller's bottom part is inserted in the hardie hole. The main function of a fuller is to either reduce the cross section of the job or to perform the necking operation.

8.2.9 Swage

Swage (Figure 8.6 (c)) is also an important smithy tool which is made in pairs, i.e., top and bottom swages. The material of swages is also high carbon steel. Bottom swage is fitted into the hardie hole, whereas the top swage is held in a tong. The main application of swage is to provide round shape to the workpiece.

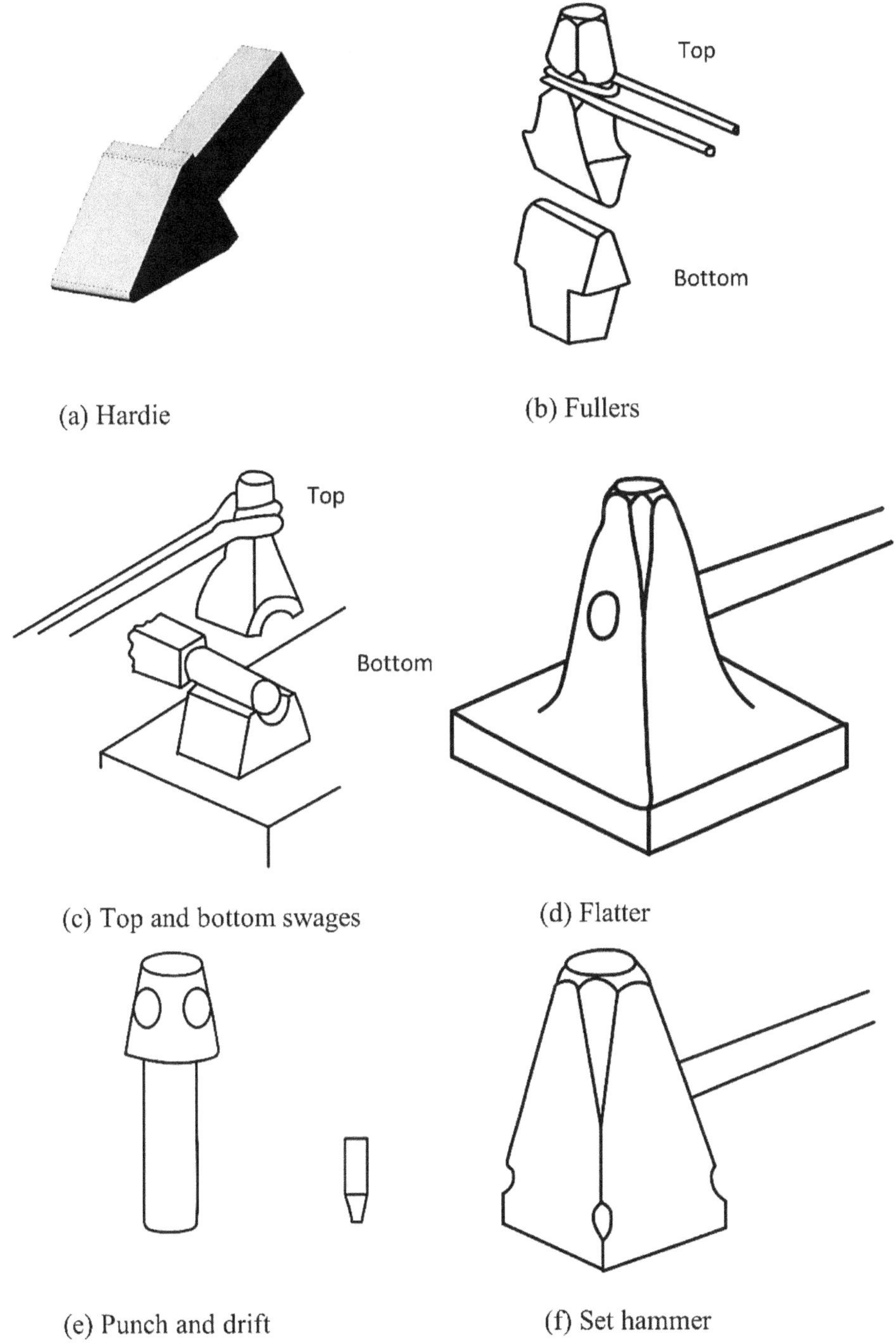

(a) Hardie

(b) Fullers

(c) Top and bottom swages

(d) Flatter

(e) Punch and drift

(f) Set hammer

Figure 8.6 Smithy tools and equipment

8.2.10 Flatter

Flatter (Figure 8.6 (d)) or smoother as it is normally called has a flat surface on one side whereas its other side has a handle. Flatters are also made of high carbon steel. The main application of a flatter is to smoothen or flatten the surfaces of a workpiece.

8.2.11 Punch and drift

Punch and drift (Figure 8.6 (e)) are available in different sizes and shapes and are selected based on the job requirements. They are also made of high carbon steel. Their main application is for punching operation and making holes. Drift is a large sized punch, which is used for enraging of holes made by the punch.

8.2.12 Set hammer

Set hammer (Figure 8.6 (f)) is also made of high carbon steel but used for flattening workpieces primarily for lighter work as compared to the flatter. Another use of a set hammer is for shoulder work, where the use of flatter is not very convenient.

8.3 SMITHY OPERATIONS

A number of operations are performed in a smithy shop, which are mentioned in the following paragraphs and with their illustrations given in Figure 8.7.

8.3.1 Drawing

Drawing (Figure 8.7 (a)) operation is used for forcefully elongating the workpiece's cross section area. This requires the application of force to be perpendicular to the cross-section area of the job with the help of tools like flatter and set hammer.

8.3.2 Up-Setting

Up-setting (Figure 8.7 (b)) is used to increase the cross-sectional area of a workpiece which however also reduces length of the workpiece. In this operation, only specified portion of the part, for which the cross-sectional area is to be increased is heated and brought to the plastic stage followed by application of force with suitable tools.

8.3.3 Fullering

Fullering (Figure 8.7 (c)) is used for reducing the cross-section area of a workpiece primarily to give it the desired shape. Fuller is the tool used in this operation, which has already been discussed in the previous section.

8.3.4 Edging

Edging (Figure 8.7 (d)) operation is used to give desired shape to the workpiece with the help of two halves of a die. Striking of the die halves with the workpiece inserted in between helps provide desired shape to the workpiece as per the die shape. The edging operation is normally performed in multiple stages to bring the workpiece to the desired shape.

8.3.5 Bending

Bending (Figure 8.7 (e)) operation is used for providing an angular bend to the workpiece. The workpiece for bending operation is generally of shapes like, sheets and rods.

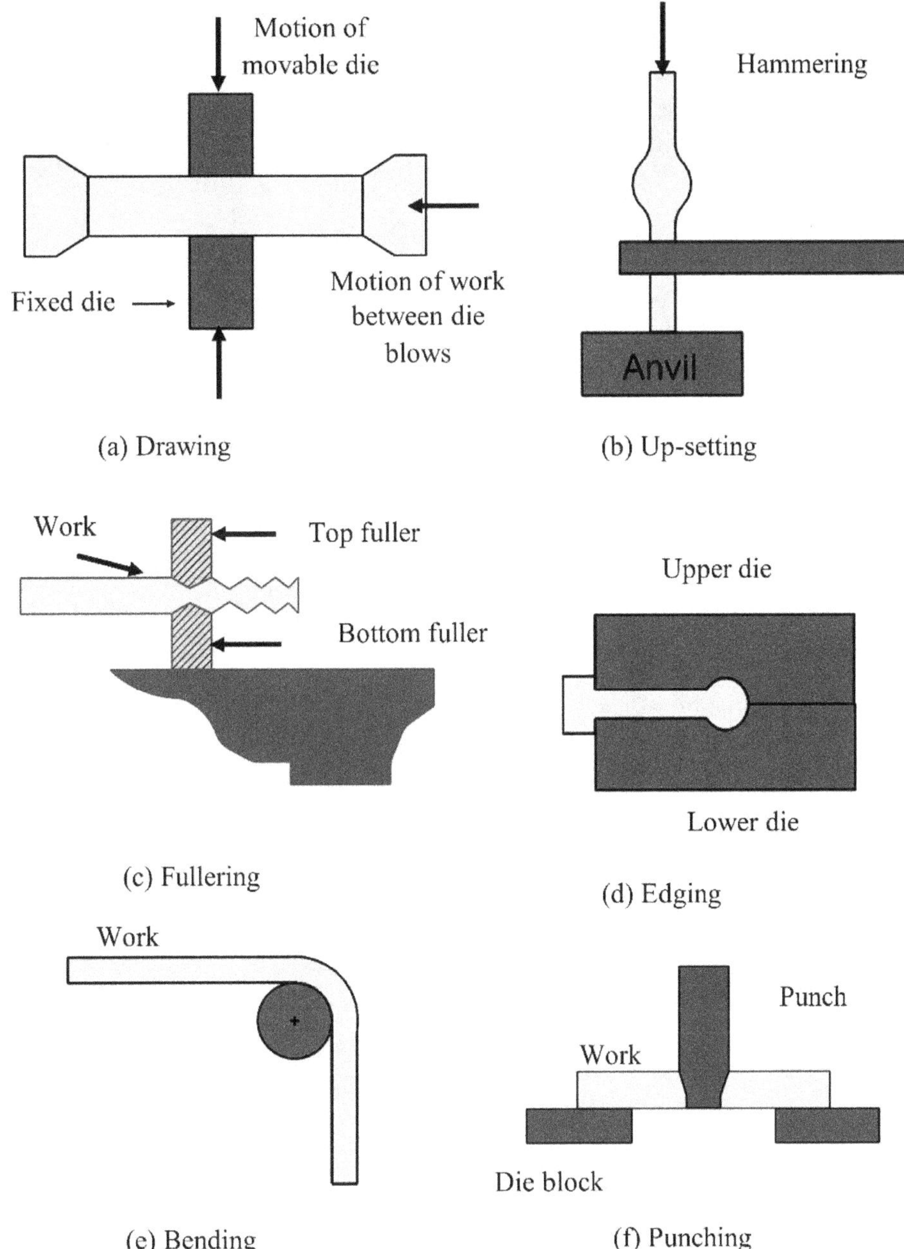

Figure 8.7 Smithy operations

8.3.6 Punching

Punching operation is used for producing holes in a metallic workpiece. The hot metal workpiece is placed over a hollow cylindrical die to allow use of punch as shown in the Figure 8.7 (f).

8.3.7 Flattening and setting down

Flattening (Figure 8.8 (a)-(b)) is used for evening out hammering marks on a job to obtain a flat surface. This operation is done with the help of a flatter or a setting hammer.

8.3.8 Swaging

Swaging (Figure 8.8 (c)) operation is used to get a workpiece reduced and finished to the desired size and shape Generally this operation is used to produce round or hexagonal shapes. For smaller workpieces, top and bottom swages are used, whereas for large workpieces a swage block can be used.

8.3.9 Cutting

Generally, in many workshop practices, cutting process involves cutting of a metal rod or a plate into two pieces with the help of a hammers and chisels. However, in the smithy and forging shop, cutting of the workpiece is done by bringing the workpiece to the plastic stage by heating and then using chisels and hammers to cut it.

8.3.10 Forge welding

Forge welding (Figure 8.8 (d)) is used for joining of two metal workpieces by first bringing them in red hot conditions followed by their pressing and hammering. Since this operation is normally done in the smithy or forging shop, the name 'forge welding' is given to this welding process.

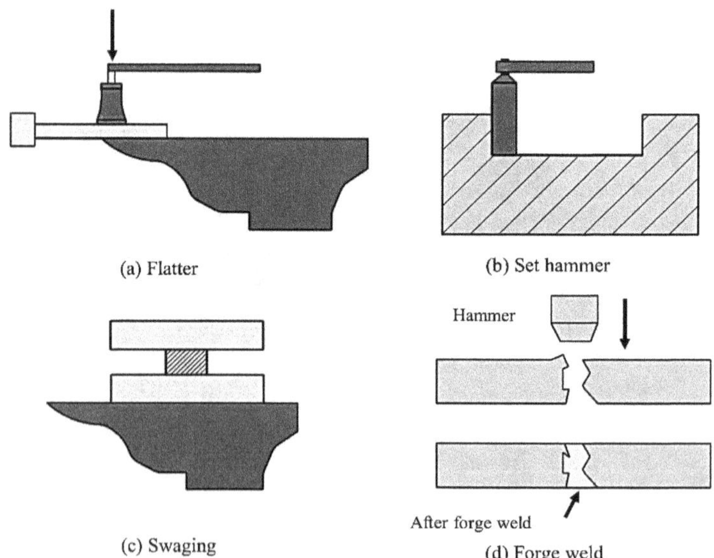

Figure 8.8 Smithy operations: flattening, set hammering, swaging and forge welding

8.4 SAFETY PRECAUTIONS

Following safety precautions are to be followed in smithy and forging shop.

i. Hold the hot work downwards close to the ground, while transferring from the hearth to the anvil. This significantly reduces the risk of burning caused due to accidental collisions with others.

ii. Always use correct sized and appropriate type of tong suited to the job shape and size. The tong should be able to hold the workpiece securely to prevent it from falling out of the tong or bouncing from repeated hammer blows.

iii. Correct sized and type of hammer should be selected keeping in view the workpiece. Only the minimum required force should be used for hammering. It should be ensured that the flat surface of the hammer strikes properly on the workpiece surface.

iv. The anvil should always be free from moisture and grease while in use.

v. Face shields or protective eyeglasses should be used when hammering hot metal.

vi. Gloves should be worn to protect hands when handling hot metal.

vii. Steel-toed shoes should be worn instead of normal shoes.

viii. This is to be ensured that the hammers are fitted with tight and wedged handles.

PRACTICE NO. 8.1

Job: To make a square shape bar from a given round rod using hand forging operation.

Objectives: Learn by practice smithy operations, such as cutting of specimen, marking, measuring, edge preparation, straightening, making right angles, and forging alongwith their respective tool handling techniques for making a square shape bar.

Machines, equipment, and tools required: Smith's hearth, anvil, ball-peen hammers, flatters, swage block, half round tongs, pickup tongs, cold chisel. marking chalk, dot punch, hand hacksaw, bench vice, flat file, square file, round file.

Material required: Round bar of mild steel or wrought iron

Schematic of the job:

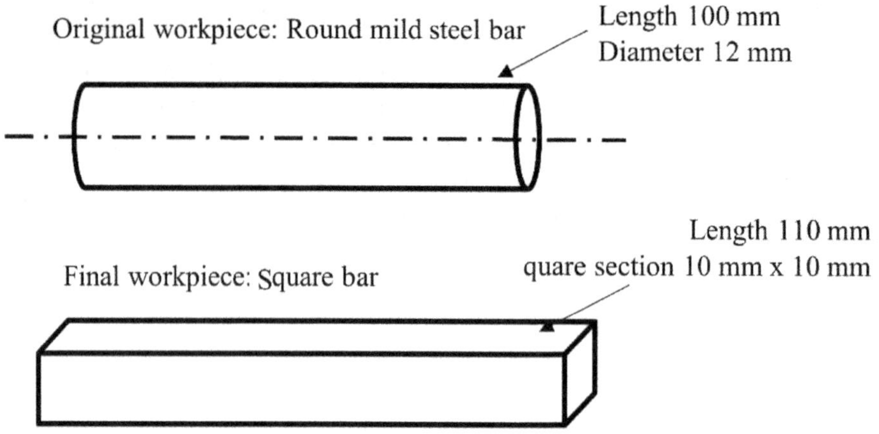

Figure P8.1 Round shaped workpiece and square shaped job

Procedure:

i. Take a round bar of mild steel or wrought iron as per the dimensions given in Figure 8.1.

ii. Ignite the open-hearth furnace and wait till the furnace is fully functional for heating.

iii. Place the round bar in the furnace and wait till the round bar becomes red hot as it comes to the plastic stage.

iv. Take the workpiece out with the help of a round hollow tong and support it on the anvil.

v. Strike the workpiece at suitable angles and force with the help of a sledgehammer to change its shape from circular to rectangular till the workpiece stays red. The workpiece may again be heated in the furnace if required. The length of the workpiece increases at the expense of its diameter.

vi. Lastly, put the job prepared in the water tank to cool it.

vii. Punch your name and roll number on the job prepared with the help of punch and hammer.

viii. Fill out the workshop practice response sheet given at the end of Chapter 2 and answer the questions provided therein. Get the feedback of the instructor on the job prepared by you.

PRACTICE NO. 8.2

Job: Make an L shaped hook from a given square bar following hand forging (smithy) operations.

Objectives: Learn by practice various smith forging operations for making a L shape metallic workpiece.

Machines, equipment, and tools required: Smith's hearth, anvil, ball-peen hammers, flatters, swage block, half round tongs, pickup tongs, cold chisel.

Material required: Rectangular bar of 110 mm x 10 mm x 10 mm size made of mild steel or wrought iron.

Schematic of the job:

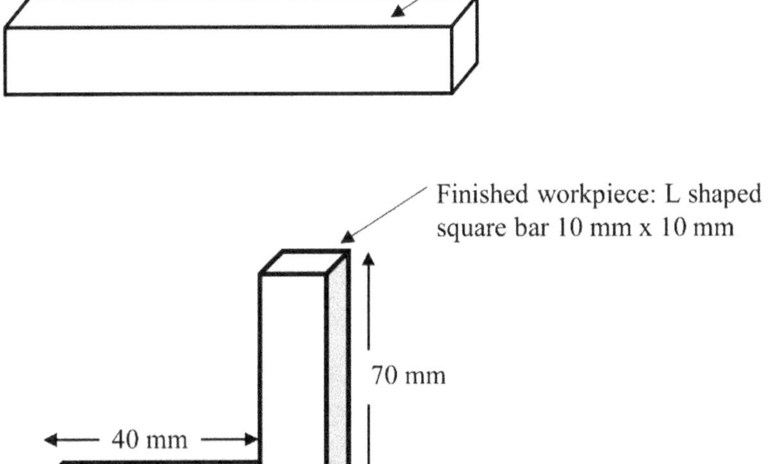

Figure P8.2 L shaped hook using a square bar

Procedure:

i. Take a rectangular bar of mild steel as per the dimensions given in Figure P8.2.

ii. Ignite the furnace and wait till the furnace is fully functional for heating.

iii. Place the rectangular bar in the furnace and wait till it becomes red hot.

iv. Take the workpiece out of the furnace with the help of square hollow tong.

v. Support it on the anvil and make an L shape with hammering by frequently changing positions of the workpiece.

vi. Put the workpiece in the water tank to cool it.

vii. Punch your name or roll number on the job prepared.

viii. Fill out the workshop practice response sheet given at the end of chapter 2 and answer the questions provided therein. Get the feedback of the instructor on the job prepared by you.

PRACTICE NO. 8.3

Job: To give an octagonal shape to a rectangular piece of mild steel.

Objectives: Learn by practice various smithy and forging operations for making an octagonal shaped metallic workpiece.

Machines, equipment, and tools required: Smith's hearth, anvil, ball-peen hammer, flatters, swage block, tongs.

Material required: Mild steel rectangular bar.

Schematic of the job:

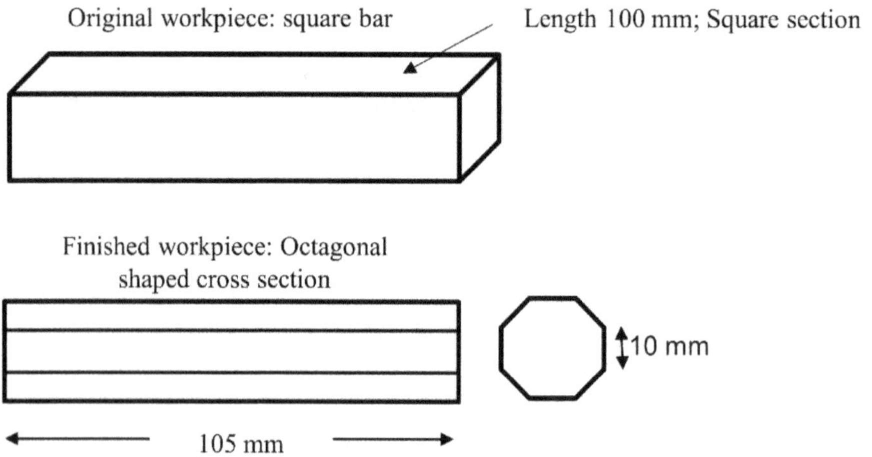

Figure P8.3 Square bar and octagonal shaped bar

Procedure

i. Take a rectangular bar of mild steel or wrought iron of the size as shown in Figure P8.3.

ii. Ignite the furnace and wait till the furnace is fully functional for heating.

iii. Place the rectangular bar in the furnace and wait till it becomes red hot as it comes to the plastic stage.

iv. Take the workpiece out of the furnace with help of a tong.

v. Support it on the anvil and hammer it intermittently frequently changing its angular position to make it octagonal shape.

vi. Put the workpiece in the water tank to cool it.

vii. Punch your name and roll number on the job prepared with the help of punch.

viii. Fill out the workshop practice response sheet given at the end of chapter 2 and answer the questions provided therein. Get the feedback of the instructor on the job prepared by you.

PRACTICE NO. NO. 8.4

Job: To convert an octagonal shaped piece of mild steel into chisel.

Objectives: Learn by practice various smithy and forging operations for making a chisel from an octagonal shaped metallic bar.

Machines, equipment, and tools required: Smith's forge, anvil, ball-peen hammers, flatters, swage block, tongs.

Material required: Mild steel octagonal bar.

Schematic of the job:

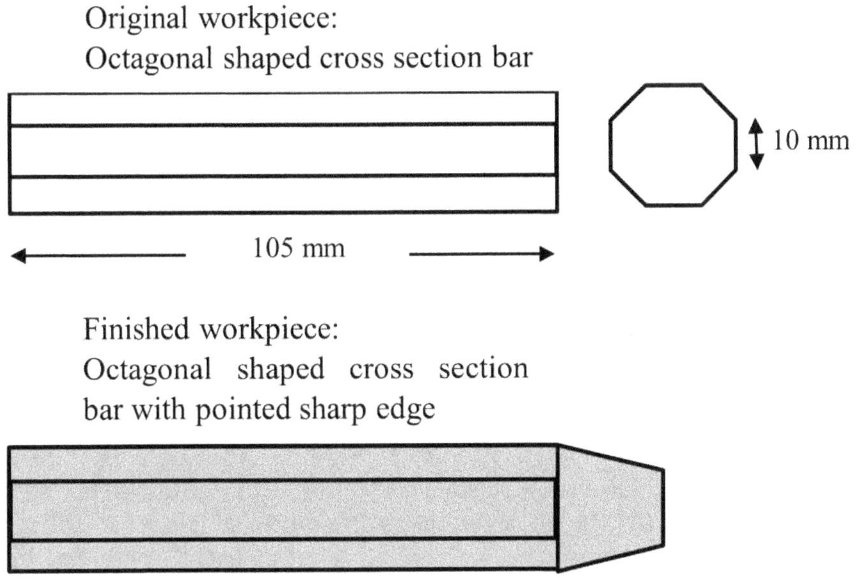

Figure P8.4 A chisel made from an octagonal shaped bar

Procedure

i. Take an octagonal shaped bar of mild steel of the size as shown in Figure P8.4.

ii. Ignite the furnace and wait till it becomes fully functional for heating.

iii. Place the octagonal shaped bar in the furnace and wait till the bar becomes red hot.

iv. Place the workpiece on an anvil with the help of a tong and start hammering to make it of taper shape on its one end.

v. Take the workpiece out of the furnace with help of round hollow tong.

vi. Put the workpiece in the water tank to cool it.

vii. Punch your name or roll number on the job prepared with the help of a punch.

viii. Fill out the workshop practice response sheet given at the end of Chapter 2 and answer the questions provided therein. Get feedback of the instructor on the job prepared by you.

Chapter 9

MACHINE SHOP

9.1 INTRODUCTION

In a machine shop the material is removed from the workpiece using a shearing process with the help of machines and cutting tools. The machine shop consists of a wide range of machine tools. The most common conventional machine tools are lathes, shapers, drilling machines, milling machines, grinding machines, planers, and slotters. Nowadays, machine shops also have the latest machinery and equipment, such as Computer Numeric Control (CNC) machines. Modern machine shops also have non-conventional machine tools, such as Electric Discharge Machines (EDM), Ultrasonic Machines (USM) and Abrasive Jet Machines (AJM). However, manually controlled lathe machine is the oldest, considered to be the mother of all machines and is indispensable. Therefore, most of the discussion in this chapter is about the lathe machine, lathe operations and cutting tools.

9.2 LATHE

A lathe machine can be used for multiple purposes to carry out a variety of machining operations. The basic principle in a lathe machine is that the work is securely placed in a fixture which is attached to the main spindle of the lathe machine, whereas the tool is provided movement along a desired path to perform the cutting operation. The principle of lathe machine is depicted with the help of Figure 9.1

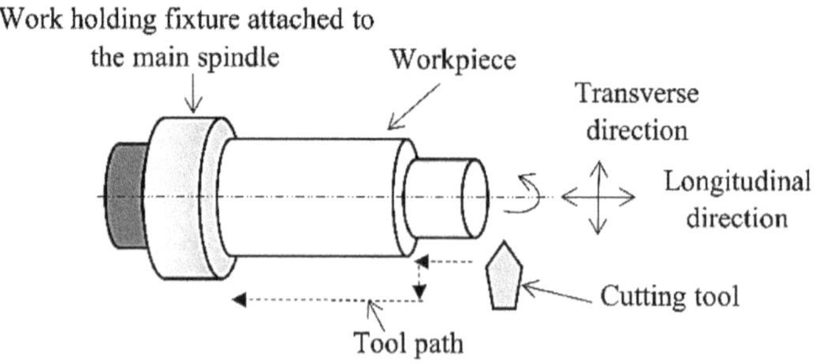

Figure 9.1: Basic principle of working of lathe machine

The procedure of working on the lathe machine is discussed in this paragraph. The work is clamped on the work holding device of a lathe machine. The work holding device, such as a chuck or a face plate, securely holds the workpiece yet allows its rotational movement. The workpiece is properly centered in such a way that there is no eccentricity during the rotational movement of the workpiece. The cutting tool, which is selected keeping in view the operation to be done is fixed on the tool post. The tool post can be moved in two directions: (i) cross direction, which is perpendicular to the axis of the spindle, and (ii) longitudinal direction, which is parallel to the axis of the spindle. The tool path is thus controlled by controlling the movement of the carriage in both directions. The normal direction of rotation of the spindle during the turning operation is the counterclockwise direction as viewed by the operator looking towards the chuck.

9.2.1 Parts of a lathe machine

A lathe machine comprises several parts, which are discussed in the following paragraphs. Figure 9.2 shows a schematic of a lathe machine and its major parts.

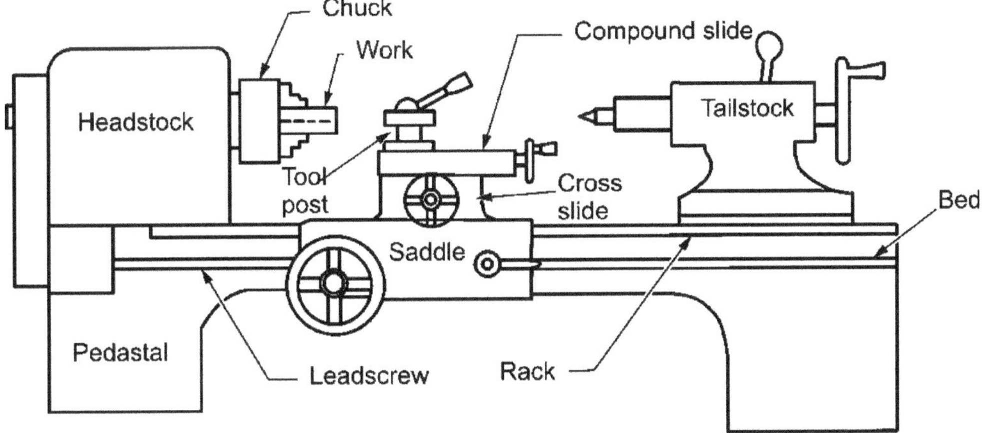

Figure 9.2 Schematic showing parts of a lathe machine

i. *Bed* is an essential part of a lathe machine which is responsible for providing a strong and rigid platform to various components of the lathe, most of which are moving parts. The application of cutting forces causes a lot of stress on the structure of a lathe, therefore, the lathe bed should be strong enough and capable of absorbing the vibrations. The lathe bed is therefore made of cast iron due to its high strength and vibration resistance properties.

    The primary function of the *lathe bed* is to provide rigid support to the parts of the lathe machine and linear and rotary movement to several parts, such as carriage and spindle.

ii. *Spindle* is the primary moving part of a lathe machine is securely placed in the bearings fitted on its head stock. Spindle of a lathe machine, which is also called the main spindle provides rotational movement to the part through the chuck.

iii. *Head stock* comprises components that are used to hold the moving parts connecting the motor with the main spindle of the lathe machine. The main spindle, which is attached to the chuck is also held by it. The spindle and hence the job can be rotated with the help of a mechanism provided in the machine. The rotational speed of the spindle can be varied with the help of either a cone pulley or a gear transmission.

iv. *Tail stock* is used to provide the required support to the workpiece, especially those which are of long dimensions. The support to the workpieces comes from the right side with the help of a dead centre mounted on the tool stock. The dead centre is a sharply pointed shaft which comes inside a blind hole provided in the workpiece for supporting the workpiece during its rotational movement. The tail stock spindle has an internal Morse taper to receive the dead center that supports the work. A secondary function of the tail stock is to perform the machining operations, like drilling, reaming, and tapping by holding the cutting tools, such as drills, reamers, and taps.

v. *Carriage or saddle* is used to control the movement of the cutting tool in transverse and longitudinal directions. Therefore, assembly of the carriage comprises two slides, a cross slide for the transverse motion and an apron for the longitudinal motion. The cross-slide helps the cutting tool to move perpendicular to the axis of the spindle. The transverse motion is used for operations like facing and grooving. The longitudinal motion is required for plain turning along the axis of the spindle.

vi. *Compound slide* is used for supporting the tool post and its movement in the angular direction to get tapered surfaces.

vii. *Tool post* is used for holding and positioning the tool holder and the tool. It provides a mechanism for adjusting the position of the tool besides the provision to hold multiple tools, however, only one tool can be engaged for cutting operation.

viii. *Lead screw* comprises a long screw which is adequately supported from the ends for controlling the movement of the carriage. Generally, acme or square threads are used for lead screw. The leadscrew is located on the front side of the lathe bed starting from the tail stock up to the head stock. The leadscrew is used for automatic feeding of the carriage by engaging it with the carriage. Furthermore, the leadscrew can also control the relative movement of the carriage and the spindle, which is needed for machining of screw threads.

ix. *Dead center* is positioned in the tail stock spindle and is used to support long jobs from the face opposite to the chuck. The dead center does not rotate with the work during the turning operation.

x. *Live center* lies in the head stock spindle. However, it revolves with the work during the turning operations.

9.2.2 Work holding devices

Three jaw chuck, four jaw chuck and face plate are the commonly used work holding devices in a lathe, which are discussed below.

i. *Three jaw chuck* (Figure 9.3 (a)) has three jaws which can be closed with the help of a tool to tightly hold the work for carrying out the lathe operations. The jaws are opened to release the part after machining work is complete. The jaws have self-centering mechanism to locate the job at the center of the spindle. Workpieces like round bars and hexagonal rods can be held in the chuck.

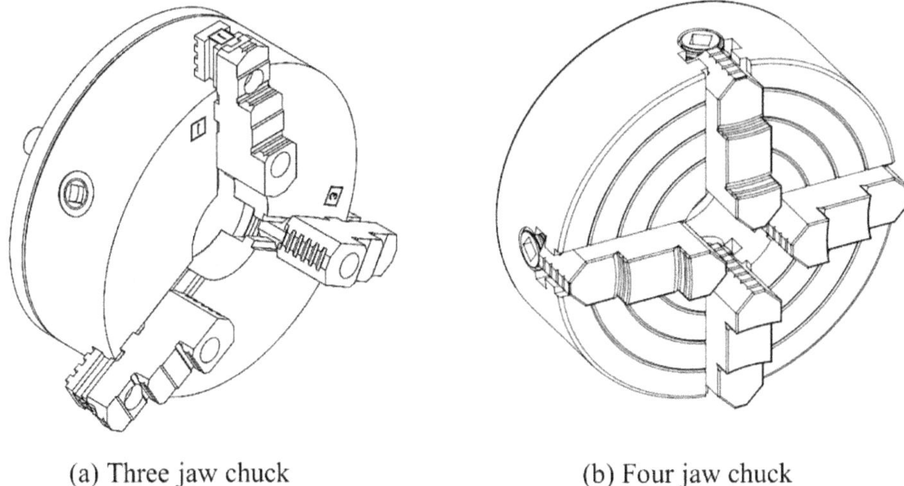

(a) Three jaw chuck   (b) Four jaw chuck

Figure 9.3 Workpiece holding devices

ii. *Four jaw chuck* (Figure 9.3 (b)), has four jaws instead of three to provide more gripping power, higher accuracy and easily hold square workpieces.

iii. *Face plate* is a circular plate of large diameter, which is also used as a work holding device. The main application of a face plate is to securely hold a workpiece that cannot be held in a chuck.

9.3 CUTTING PARAMETERS

The most important cutting parameters for machining on a lathe machine are discussed in the following paragraphs.

i. *Cutting speed* is defined as the speed of the workpiece surface at the point of contact with the tool. The cutting speed may be specified either in meters per minute (m/min) or rotations per minute (rpm). The cutting speed for a turning operation is selected after considering factors, such as work piece material, feed, depth of cut, type of operation and other cutting conditions. The cutting speed can be easily determined by using the Equation (9.1).

$$CuttingSpeed = \frac{\pi \times D \times N}{1000} \quad \text{(m/min)} \qquad (9.1)$$

where, D is the diameter of the workpiece in mm, and N is the spindle revolutions per minute.

ii. *Feed* is described as the travel speed of the cutting tool in the direction of cut. This may either be described in terms of mm per revolution or mm per minute. The feed rate is determined keeping in view the factors of depth of cut and required surface finish of the work.

iii. *Depth of cut* is the extent of insertion of the cutting tool inside the surface of the workpiece. Depth of cut is decided keeping in view the surface finish required and the power available with the machine tool.

9.4 TOOL MATERIALS

The cutting tools used in various lathe operation are generally made of carbon steel or high-speed steel. The high-speed steel (HSS) tools have mainly three alloying elements, 18 % tungsten, 4 % chromium and 1 % vanadium apart from 0.7 % carbon, the rest being iron. Sometimes. cobalt to the extent of about 5% to 10% is also added to improve the heat resisting properties of the HSS tool.

A wide variety of tools with different materials are available for lathe. For example, carbide tips are fixed in tool holders to make tools which can be used for very high-speed cutting. Details of other tool materials can be found in books on machining processes and tools.

9.5 CUTTING TOOL

A number of cutting tools (Figure 9.4) are available each suiting for specific machining operation to be carried out on a lathe machine. The most used tool for lathe machines is the single point cutting tool. which is generally made from a hardened steel bar, such as high-speed steel (HSS) or high carbon steel by providing its one end properly shaped and sharpened to make a cutting edge. Another way of making a single point cutting tool is by making use of a hardened ceramic tool tip, which is securely fixed to one end of a rectangular steel bar in such a way that it can perform the cutting action. The hardened tool tip is mounted with the help of screws or by using brazing operation. The tip of hard material is generally made of cemented carbide.

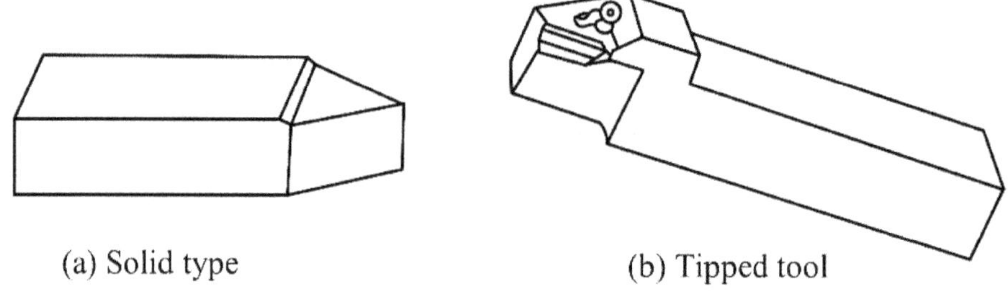

(a) Solid type   (b) Tipped tool

Figure 9.4 Single point cutting tools: (a) solid type, and (b) tipped tool

9.5.1 Tool geometry

The geometry of a single point cutting tool, which is the most used in the lathe operations is shown in Figure 9.5 with its major features indicated. The main features of a single point cutting tool are nose, rake face, flank, heel and shank. Features of a typical single point cutting tool are described in the following paragraphs.

i.   *Shank* is the main supporting region of a single point cutting tool which is inserted in the tool holder to hold the tool.

ii.   *Cutting edge* is the sharp edge of the tool which is responsible for carrying out the shearing operation to remove the materials from the workpiece. A single point cutting tool has two cutting edges, namely main (or side) cutting edge, and the secondary (or end) cutting edge.

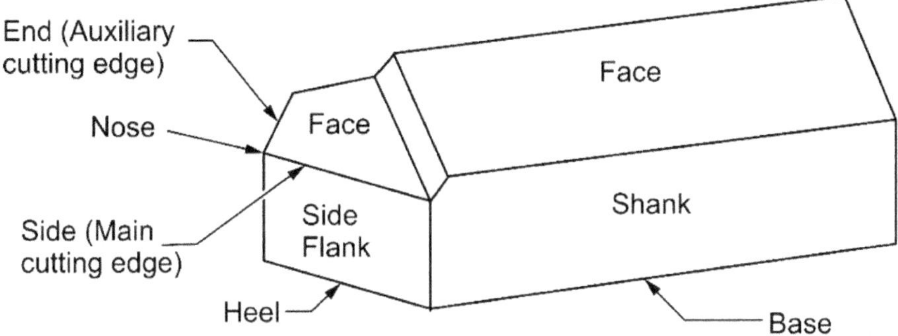

Figure 9.5 Single point cutting tool geometry

iii.   *Flank* is the tool's face, which is adjacent to the cutting edge. The major flank is just below and adjacent to the main (or side) cutting edge and the minor flank is just below and adjacent to the secondary (or end) cutting edge.

iv.   *Base* is a face of the cutting tool, which lies just opposite to the top face of the shank.

v.   *Face* is the top surface of the tool on which the chips slide.

vi.   *Nose* is at the intersection of the main and the secondary cutting edges. The nose is not perfectly sharp but is slightly rounded. The roundness is defined with the help of the nose radius. The nose radius increases the life of the tool and is helpful to provide the workpiece surface better surface finish. However, a very large nose radius decreases the surface finish, and therefore such a tool is considered blunt.

vii.   *Heel* is at the intersection of the base and the flank of the tool and is curved in shape.

9.6 LATHE OPERATIONS

The important lathe operations are discussed in the following paragraphs.

i.   *Facing* (Figure 9.6 (a)) operation is used to machine ends of a workpiece. In this operation, the cutting tool is moved in the direction perpendicular to the spindle axis with the help of the machine's cross slide.

ii.   *Parting off* (Figure 9.6 (b)) operation is used for cutting the workpiece into two parts using a parting off tool. This operation also requires feed to be given in the transverse direction, just like the facing operation.

iii.   *Drilling* (Figure 9.6 (c)) operation is used to machine holes which are located along the axis of the spindle. Drilling operation on a lathe machine is conducted by holding a drill tool in the tail stock, whereas rotation is provided to the workpiece held in the chuck.

iv.   *Turning* operation is used to remove excess material from the workpiece to produce a cylindrical surface of the desired diameter and length. In this operation, the tool is moved in the longitudinal direction by providing a suitable depth of cut. Plain turning, step turning and taper turning shown in Figure 9.6 (d)-(f)) are examples of different types of turning operations.

v.   *Boring* (Figure 9.6 (g)) operation is used for increasing the size of the hole which is already machined using the drilling operation. This operation can also be used when a correct size drill is not available or for obtaining higher accuracy. The boring operation, however, cannot be used for creating a fresh hole.

vi.   *Knurling* (Figure 9.6 (h)) operation is used to create a specific pattern on the surface of a turned component. The knurling tool is held securely on the tool post of the lathe machine, whereas the workpiece is rotated. This operation requires using a knurling tool, which has rollers having the same pattern that is to be created on the workpiece surface. The resulting surface thus becomes rough and easy to grip and handle.

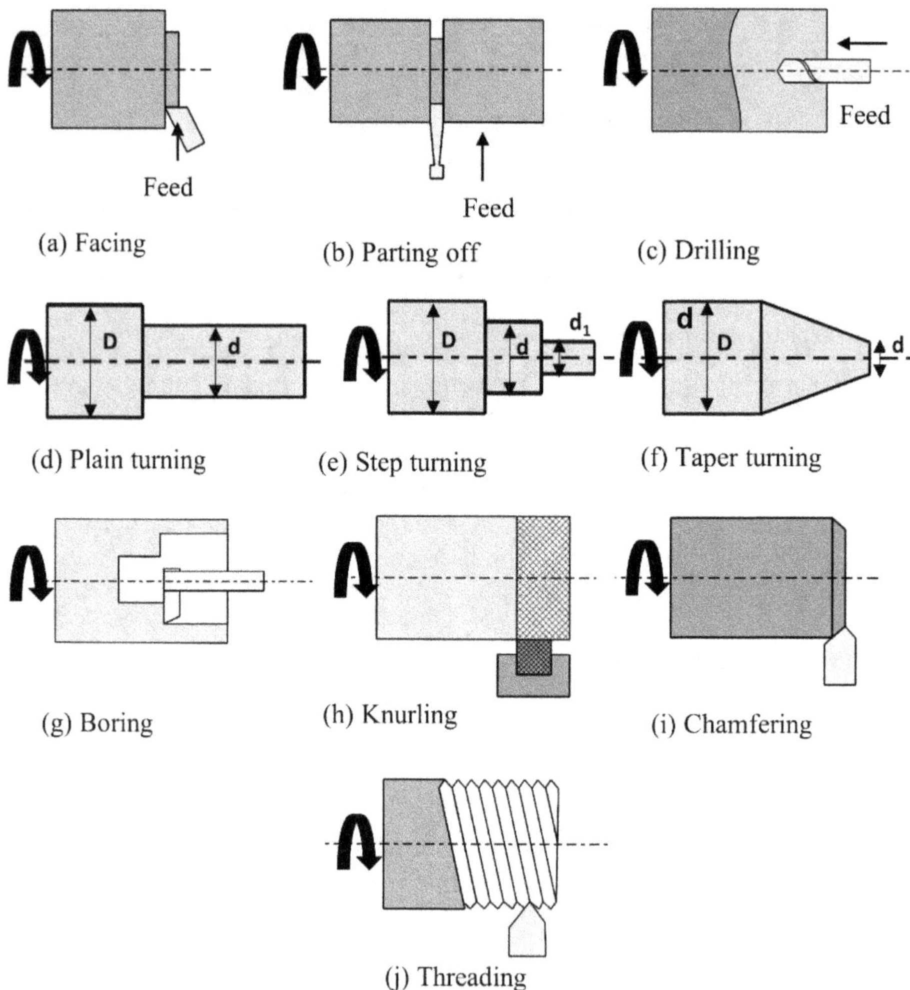

Figure 9.6 Lathe operations

vii.   *Chamfering* (Figure 9.6 (i)) operation is used to create a beveled surface at the end of a workpiece. The purpose of a chamfer is to provide a better look. Another purpose of chamfering is to enable the nut to pass freely on a threaded workpiece as the burrs on its sharp corners/edges are removed by chamfering. The chamfering operation also helps prevent damage to the sharp edges of a workpiece.

viii.   *Threading* (Figure 9.6 (j)) operation is used to create threads on a workpiece's surface, which in machining language is about the cutting of helical groove of desired shape and depth. Threads may be cut either on the internal or external cylindrical surfaces of a workpiece.

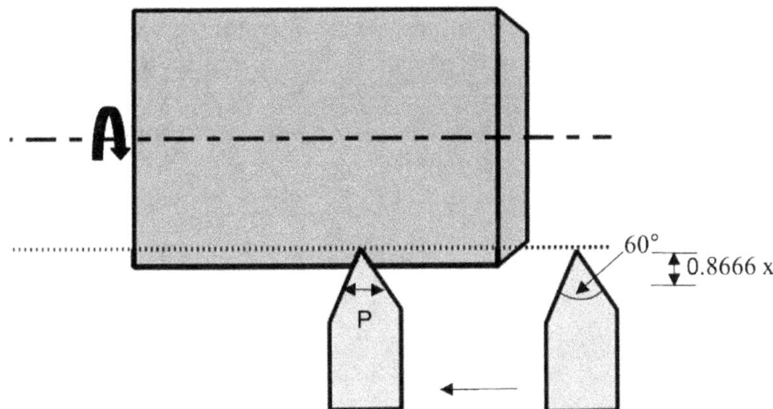

Figure 9.7 Principle of threading operation

The principle of machining of threads is depicted in Figure 9.7. The threading tool is a specially shaped cutting tool, which depends on the type of thread required. For example, a metric thread needs a cutting tool with 60-degree profile. Further, thread cutting in a lathe is performed by moving the cutting tool at a definite rate, in proportion to the rate at which the work revolves. For example, if one revolution of the workpiece also corresponds to 4 mm movement of the tool in the longitudinal direction, it means that a thread of 4 mm pitch will be created. Pitch is also mentioned as 1/TPU, where TPU is thread per inch. The depth of cut required for machining a thread of pitch P will be 0.866 x P for metric thread keeping in view the fact that the included angle of metric thread is 60 degree and cos 60° equals 0.8666.

9.7 SAFETY PRECAUTIONS

i.   Wear eye protection gear - Industrial quality safety glasses with side-shields should be worn. During machining process on a lathe machine, very small sharp and hot metal pieces are thrown away at high speeds, which can be dangerous for the eyes.

ii.   Wear tight sleeve shirts - Loose sleeves can get entangled with the rotating workpiece causing injury to the operator.

iii.   Wear shoes - Wear shoes made of leather or thick cloth to protect your feet from hot metal chips and other sharp materials.

iv.   Remove wrist watches, necklaces, chains, and other jewelry. Furthermore, if you have long hair which may get entangled in the rotating work, tie them properly. Just imagine what will happen to your face if it gets entangled in the machine by chance.

v.   Clamp the work properly in the chuck before starting the lathe. The lathe should be started at a low speed first and thereafter the speed is increased gradually.

vi.   Remove the chuck key immediately after use as it may become a projectile ready to hit the operator if it is left unattended in the chuck.

vii.   Keep your fingers clear and at a safe distance from the rotating work and the cutting tool. Hands should not be used to break away the continuous chips that may be formed.

viii.   Avoid reaching over the spinning chuck. For filing operations, the tang end of the file should be held with the left end. This is to ensure that your hand and arm are not above the spinning chuck.

ix.   Never use a file with a bare tang. The tang end should always have a handle attached to it. This is because if the tang is bear, it may cause injury to your hand.

PRACTICE NO. 9.1

Job: To prepare a job as per sketch using a lathe machine.

Objectives: Learn by practice various machining operations, namely turning, facing, taper turning, grooving, and chamfering on lathe machine for making a job of given dimensions.

Machines, equipment, and tools required: Centre lathe machine, three jaw chuck and related accessories, right hand side turning tool, grooving/parting tool, steel rule, vernier caliper and micrometer.

Material required: Round mild steel bar of diameter of 30 mm to 32 mm and length 125 mm approximately.

Schematic of the job:

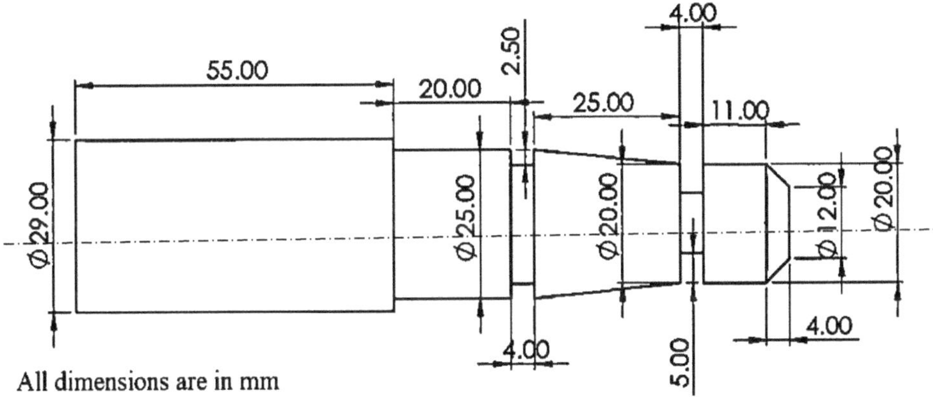

(a) Dimensions for job preparation

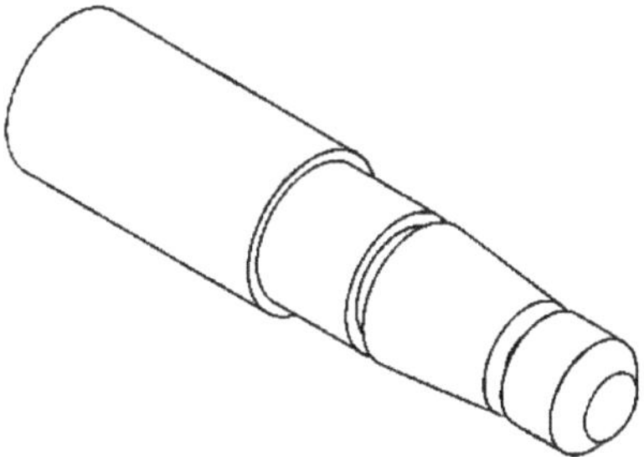

(b) Isometric view of prepared job

Figure P9.1 Schematic of the job to be prepared (a) dimensions of the job, (b) isometric view of the prepared job.

Procedure:

i.   Hold and tighten the cylindrical workpiece in the chuck leaving sufficient length for the machining operations.

ii.   Hold the cutting tool in the tool holder and fit it on the tool post.

iii.   Perform facing operation on the right-hand face of the workpiece.

iv.   Mount a plain turning tool on the tool-post and perform the plain turning operation with a diameter of 29 mm over complete length of the workpiece.

v.   Perform the plain turning for a diameter of 25 mm up-to the required length as per the part description given in Figure P9.1.

vi.   The plain turning for a diameter of 15 mm up-to the length of 19 mm from the right end of the workpiece.

vii.   Perform the taper turning operation by swiveling the compound slide as per the drawing of the part starting from diameter of 15 mm going up to a diameter of 25 mm. The compound lower rest surface is graduated with degrees and the compound slide can be swiveled to any angle with the axis of the workpiece.

viii.   Demount the plain turning tool and replace it with the left-hand turning tool. Machine the shoulder at the end of the 25 mm diameter step of the job.

ix.   Demount the side turning tool and mount a grooving/parting tool in the tool holder. First perform grooving operation with a 5 mm deep cut at the end of diameter 15 mm step keeping first step length of 15 mm from the right end of the job.

x.   Similarly machine the second groove with 2.5 mm deep cut of 4 mm width at the end of the taper. This will bring diameter of the groove to be 10 mm.

xi.   Perform the chamfering operation, which is a very small taper at the rightmost end of the job.

xii.   Fill in the workshop practice response sheet given at the end of Chapter 2 and answer the questions provided therein. Get feedback of the instructor on the job prepared by you.

PRACTICE NO. 9.2

Job: Machine a job as per the sketch using a lathe machine.

Objectives: Learn by practice various machining operations, namely turning, facing, taper turning, grooving, chamfering, drilling, and knurling on a lathe machine for making a job of given dimensions.

Machines, equipment, and tools required: Centre lathe machine, three jaw chuck and related accessories, right hand side turning tool, grooving tool, parting tool, steel rule, vernier caliper and micrometer

Material required: Round mild steel bar of approximately 30 mm diameter and length of 105 mm approximately.

Schematic of the job:

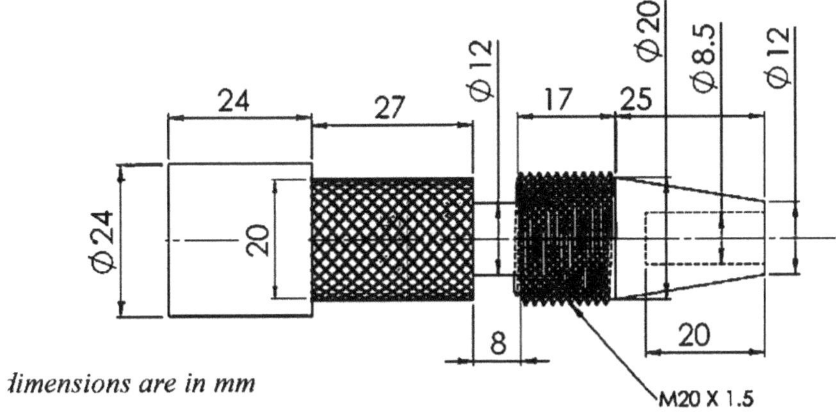

(a)   Dimensions for job preparation

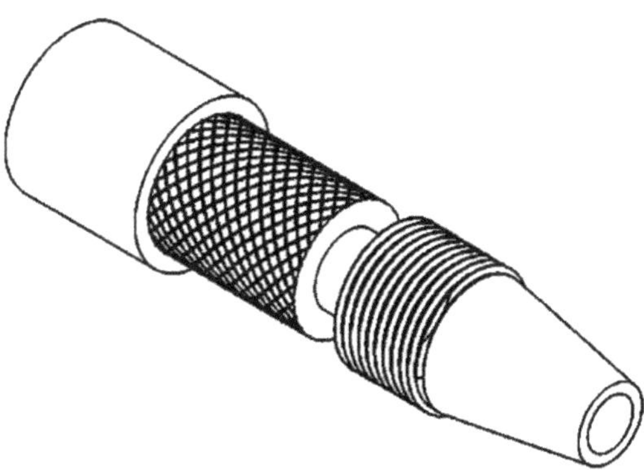

(b)   Isometric view of prepared job

Figure P9.2 Schematic of the job to be machined (a) dimensions of the job, (b) isometric view of the job prepared.

Procedure:

i. Hold and tighten the cylindrical bar in the chuck ensuring that sufficient length for operations is available.

ii. Hold the side tool in the tool holder and fix it on the tool post.

iii. Mount a single point plain turning tool on the tool post with the help of the tool post key.

iv. Perform the facing operation on the rightmost surface of the workpiece.

v. Now perform the drilling operation to machine a hole of size 25 mm (length) and 8.5 mm (diameter) as shown in the figure.

vi. Perform plain turning for a diameter of 25 mm over complete length of the workpiece as per the part drawing.

vii. Perform plain turning for a diameter of 20 mm upto the length from the left end as per the given drawing.

viii. Demount the plain turning tool and fix left hand side turning tool. Machine the shoulder at the end of 20 mm diameter.

ix. Demount the left-hand side turning tool and fix the grooving tool on the tool post. Perform the grooving operation as per the dimension given in the part drawing.

x. Perform the threading operation. Single start threads will be made on the lathe machine by adjusting the pitch value and reducing the spindle speed and automatic feed settings.

xi. The last operation is the diamond knurling at the specific place mentioned in the drawing with the help of a knurling tool.

xii. Fill in the workshop practice response sheet given at the end of Chapter 2 and answer the questions provided therein. Get feedback of the instructor on the job prepared by you.

PRACTICE NO. 9.3

Job: To perform various lathe machine operations to make a part as per given drawing.

Objectives: Learn by practice various machining operations, namely turning, facing, taper turning, grooving, and chamfering on lathe machine for making a job of given dimensions.

Machines, equipment, and tools required: Centre lathe machine, three jaw chuck and related accessories, right hand side turning tool, grooving tool, parting tool, threading tool, drill bit, boring tool, chamfering tool, steel rule, vernier caliper and micrometer.

Material required: Round mild steel bar of diameter 30 mm and length 110 mm approximately.

Schematic of the job:

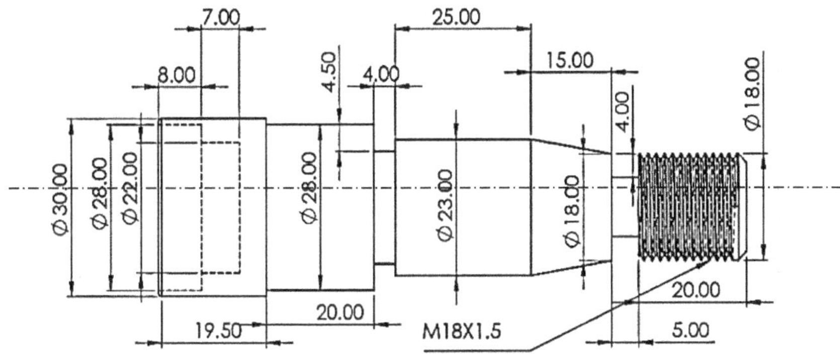

(a) Dimensions for job preparation

All dimensions are in mm

(b) Isometric view of prepared job

Figure P9.3 Schematic of the job to be prepared (a) dimensions of the job, (b) isometric view of the prepared job.

List of operations:
   i.   Facing
   ii. Center Drilling
   iii. Threading
   iv. Grooving
   v.   Taper turning
   vi. Knurling
   vii. Step turning
   viii. Drilling
   ix. Boring
   x.   Chamfering

Procedure:

i. Hold, tight and center the cylindrical workpiece in the chuck leaving sufficient length for the machining operations.

ii. Hold the cutting tool for facing in the tool holder and fit it on the tool post.

iii. Perform facing of the front face (face away from the chuck) of the workpiece. Also make central hole with a center drill on this face.

iv. Perform plain turning operation on the job for a diameter of 30 mm up to a length of 110 mm using a plain turning tool.

v. Perform the plain turning operation on the job for a diameter of 28 mm up to a length of 90 mm using a plain turning tool.

vi. Perform the plain turning operation on the job for a diameter of 23 mm up to a length of 70 mm using a plain turning tool.

vii. Perform the plain turning operation to make the first step for a diameter of 18 mm using a plain turning tool.

viii. Mount a side cutting tool and machine the unmachined areas near the end of each step.

ix. Demount the side cutting tool and mount groove or parting tool in the tool holder. Perform grooving operation to make a groove of size 5 mm (width) and 4 mm (depth) as per the requirement at the end of diameter 18 mm step keeping first step length of 20 mm.

x. Perform another grooving operation with size 5 mm (width) and 4.5 mm (depth) as per the requirement at the end of diameter 23 mm, which is at 65 mm from the right end.

xi. Demount groove/parting tool and mount side turning tool to machine the end shoulder at the end of 28 mm diameter.

xii. Calculate the taper angle for taper turning using the equation mentioned below.

$$tan\theta = \frac{D - d}{2 \times L}$$

where, θ is the semi cone angle, D is big diameter, d is small diameter and L is taper length (horizontal).

xiii. Set the compound rest to the angle calculated and turn the taper in steps giving feed in the compound slide.

xiv. Fix the threading tool and calculate the change gears and set them accordingly for machining of threads.

xv. Engage the tumbler gear to give feed to the lead screw and cut the threads in steps.

xvi. Demount the threading tool and mount a knurling tool to perform the knurling operation to a length of 25 mm as shown in the part description.

xvii. Demount the workpiece from the chuck and hold the job from the reverse side.

xviii. Now mount the drilling tool and produce a hole of 22 mm in diameter and 16 mm in length by moving the tailstock.

xix. Enlarge the hole by mounting a boring tool as per the required dimensions.

xx. Perform the chamfering operation using a chamfering tool.

xxi. Remove the workpiece from the chuck and check all the dimensions using a vernier caliper and a micrometer.

xxii. Fill in the workshop practice response sheet given at the end of Chapter 2 and answer the questions provided therein. Get feedback of the instructor on the job prepared by you.

Chapter 10

SHEET METAL SHOP

10.1 INTRODUCTION

Sheet metal shop is the place in which thin metallic sheets are worked upon to make useful products. Most of the tools used in sheet metal working are simple hand tools, which are used to perform a number of operations, such as shearing (or cutting), bending and joining.

10.2 SHEET METAL MATERIALS

Different types of metallic sheets are used in sheet metal working, such as aluminum and mild steel. Table 10.1 provides names of important material used for sheet metal working along with their properties and applications.

Table 10.1 Materials for sheet metal working

Material	Properties	Applications
Black iron	Uncoated iron sheets are cheapest available for sheet metal working.	Tanks, trunks, stove pipes
Galvanized iron	Iron sheets are coated with Zinc, is resistant to corrosion.	Cabinets, trunks, buckers, pans
Copper sheets	Good corrosion resistance, excellent thermal and heat conductivity, good appearance but is costly.	Automobiles and domestic heating appliances.
Aluminum sheets	High resistance to corrosion and abrasion. Light in weight.	Kitchen ware, aeroplane fabrication
Tin plates	Iron sheets coated with tin. Resistant to corrosion and rust.	Food containers, cans
Stainless steel	Highly corrosion resistant and tougher than galvanized iron.	Kitchen ware and food handling equipment.
Brass	Light and strong but is costly	Kitchen ware and utensils

Lead	High resistance to acidic corrosion	Radiation shielding and inner lining of acid tanks
Zinc	Highly ductile	Roofing work, coating, die casting

10.3 SHEET METAL TOOLS

A number of tools are used for sheet metal working, which are discussed in the following paragraphs.

10.3.1 Trammels

Trammels are used for marking arcs and circles on metallic sheets. Figure 10.1 (a) shows a snapshot of a trammel.

10.3.2 Standard wire gauge

This is one of the most important measuring tool in the sheet metal industry. The thickness of a sheet is referred with respect to its gauge number measured with the help of a standard wire gauge (SWG). Normally, thickness of the sheet used ranges from 16 gauge down to 30 gauge. The gaps in the circumference of the gauge are used to check the gauge number of the sheet. Figure 10.1 (b) shows a snapshot of the wire gauge.

10.3.3 Bench shears

Shearing action is an important operation in sheet metal working which is done with the help of bench shears. A bench shear makes use of a lever mechanism to drive a blade which can cut sheets of about 4 mm thickness. The chopping hole provided in a bench shear can shear mild steel rods of about 10 mm diameter. Figure 10.1 (c) shows a snapshot of bench shears.

10.3.4 Snips

Snips (Figure 10.1 (d)) are basically hand shears, which may be of different lengths ranging from 200 mm to 600 mm, however, snips with a length of about 250 mm are the most common. Depending on the type of application, different types of snips, such as straight, curved, and bent are available.

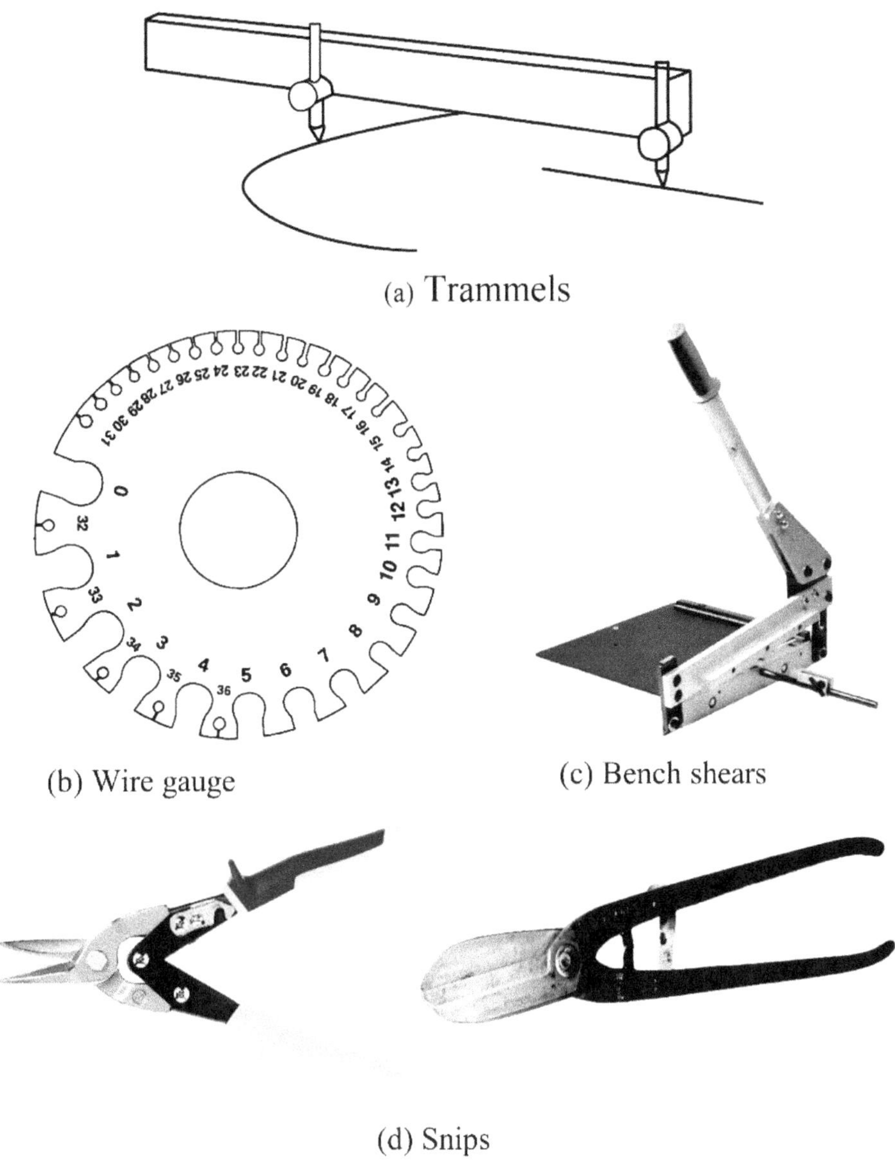

(a) Trammels

(b) Wire gauge

(c) Bench shears

(d) Snips

Figure 10.1 Commonly used sheet metal tools

10.3.5 Hammers

Hammers are used in the sheet metal shop to straighten the sheets or to give them the required shape. A wide variety of hammers to be used in sheet metal shop are available. For example, square head hammer has flat square head on one end and a circular flat end on the other. Ball peen hammer has a curved cylindrical face on one end and a flat cylindrical on the other end. Straight peen hammer has a sharp-edged peen on one end, which is parallel to the handle of the hammer. Similarly, the cross-peen hammer has a sharp-edged peen on one end, which is at right angle to the handle of the hammer. Figure 10.2 ((a)-(e)) shows snapshots of different types of hammers.

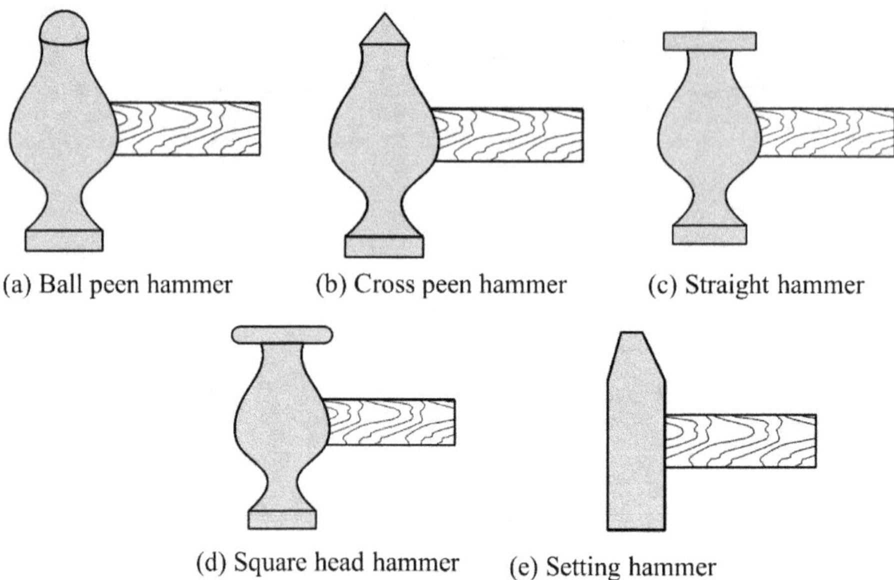

Figure 10.2 Hammers

10.3.6 Stakes

Stakes are used for forming of metal sheets in different shapes. A stake has two major features, a shank, which is provided with a specific shape and a head, which is used to hold it. Figure 10.3 shows a snapshot of stakes.

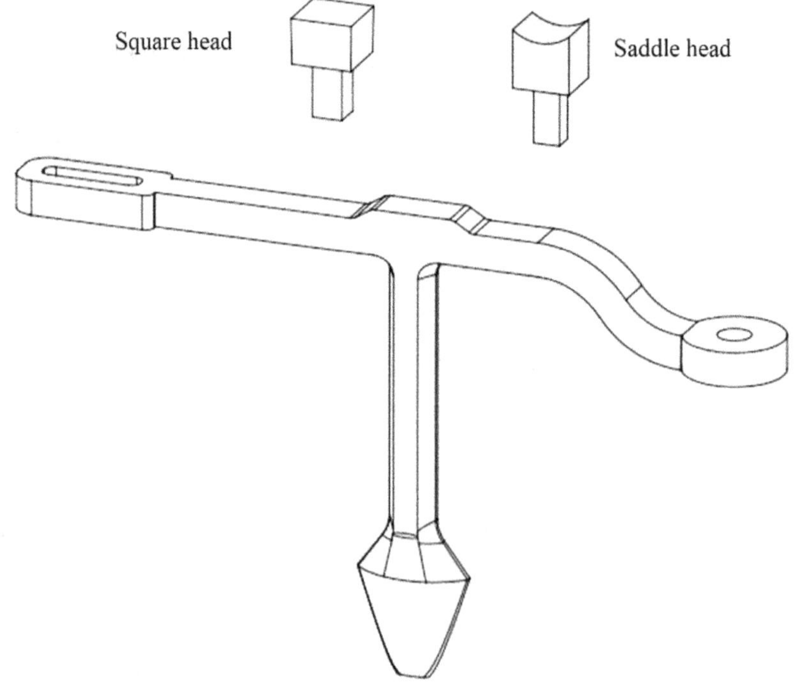

Figure 10.3 Stakes

10.4 SHEET METAL OPERATIONS

Sheet metal working operations are discussed in the following paragraphs.

10.4.1 Bending

Sheet metal jobs often need bending of the sheet at several places. Furthermore, the shape and curvature of the bends may also vary depending upon the design requirements of the part to be produced. A sheet metal may be bent by hammering with the support of a base which may be provided by a die, a rolling machine or an anvil. Figure 10.4 (a)-(e) shows snapshots of the bending operations being done with different methods.

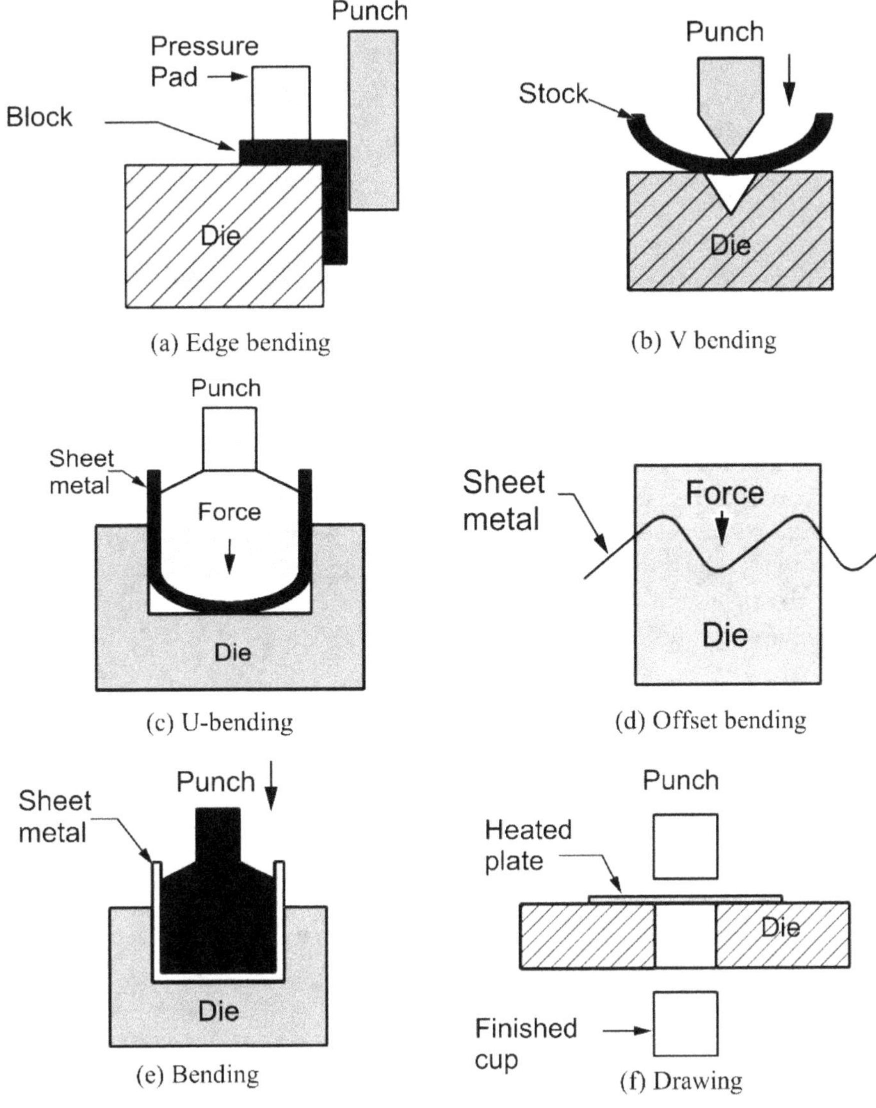

Figure 10.4 Bending and drawing operations

10.4.2 Drawing

Drawing operation (Figure 10.4 (f)) is used for producing hollow shapes from the sheet metal. The drawing operation is carried out by using a die and a punch on a suitable press. Drawing operation can further be divided into two, shallow drawing and deep drawing. As the name indicates, when the drawn length of a component is less than its width (or diameter) it is called shallow drawing. On the other hand, when the drawn length is more than the width, it is called deep drawing.

10.4.3 Soldering

Soldering, which is generally used for joining wires in electronics components can also be used for joining sheet metals. The soldering alloy is spread between the surfaces of the sheet metal to be joined. The soldering alloy solidifies after some time to make a joint. A practice on soldering is provided in chapter 6 (electrical shop), therefore it is not being discussed here to avoid repetition.

10.4.4 Embossing

Embossing (Figure 10.5 (a)) is used in the sheet metal working for creating raised surfaces or letters. However, this process causes no change in the thickness of the sheet metal. The embossing operation is done with the help of a die which has two halves to make the required impression on the sheet.

10.4.5 Shearing

Shearing operation (Figure 10.5 (b)) is the most basic operation in sheet metal working. This operation is carried out with the help of tools like bench shear, hand shear or snip. In this operation, the sheet metal workpiece is kept between the two shears of a shearing tool or two sharp edged die halves, and a force is applied on the sheet which results into shearing of the sheet. Shearing operations, such as stretch forming and deep drawing are shown in Figure 10.5 ((c)-(e)).

10.4.6 Blanking

Blanking (Figure 10.5 (f)) is a sheet metal operation in which the shearing operation is used to obtain a useful part from it. However, in this operation, the original sheet is left as a scrap. This operation makes use of a punch and a die for carrying out the operation.

10.4.7 Punching

Punching operation (Figure 10.5 (g)) is used to create a hole of desired shape in a sheet metal component. In this operation, the material that comes out of the original sheet metal workpiece is waste or scrap. Therefore, this operation may be considered as opposite of the blanking operation as far as the treatment of the scrap is concerned. This operation also makes use of a punch and a die.

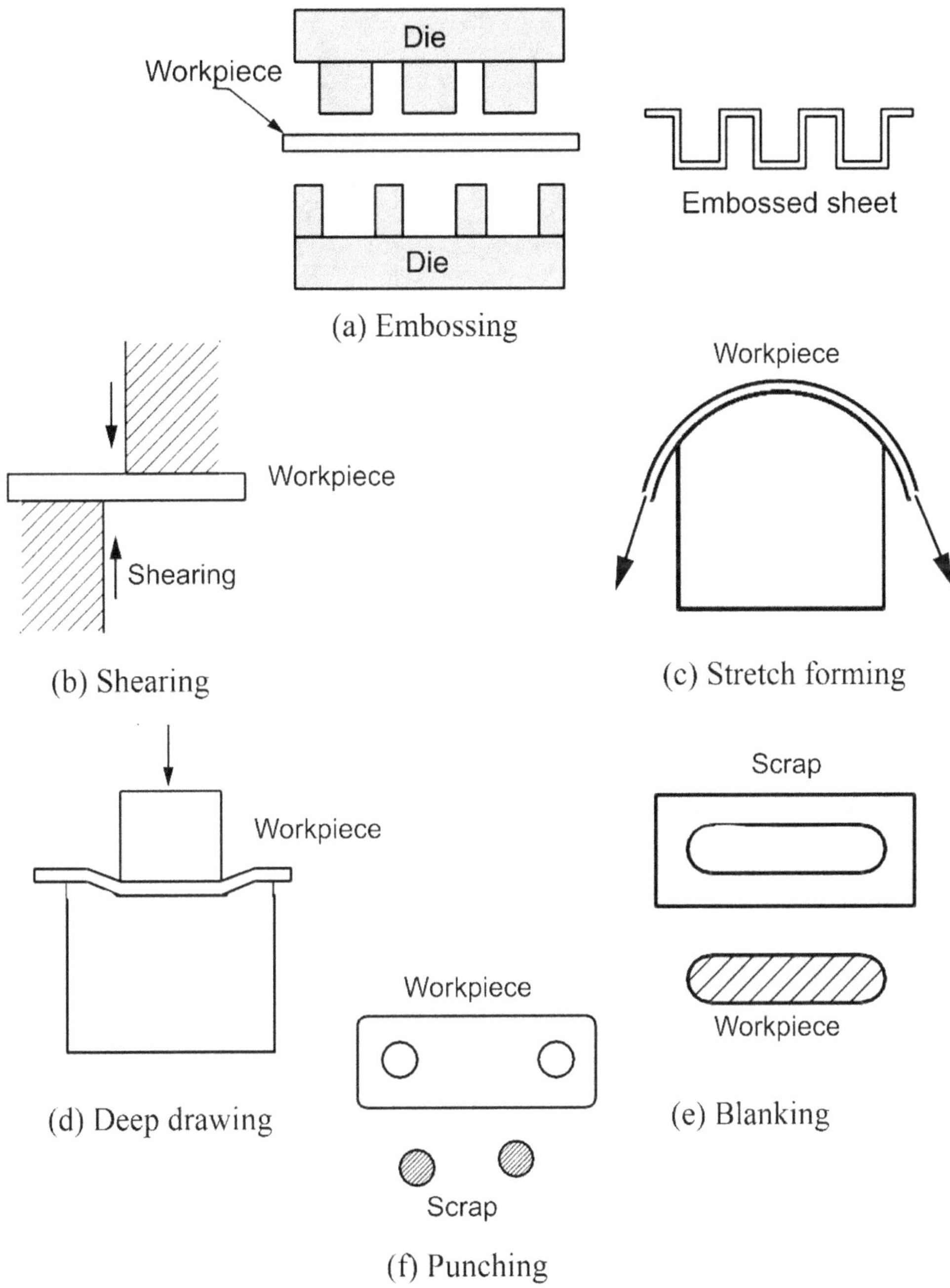

(a) Embossing

(b) Shearing

(c) Stretch forming

(d) Deep drawing

(e) Blanking

(f) Punching

Figure10.5 Embossing, shearing, blanking, and punching operations

10.4.8 Lancing

Lancing operation (Figure 10.6 (a)) in sheet metal working creates desired shaped holes in the workpiece without removal of the material from the sheet. This means that the piecing operations create a partial cut in the sheet so that the contact of the sheared material from the workpiece does not completely vanish. This operation, like punching and blanking, also needs a die and a punch which are of the same shape as that of the shear to be created.

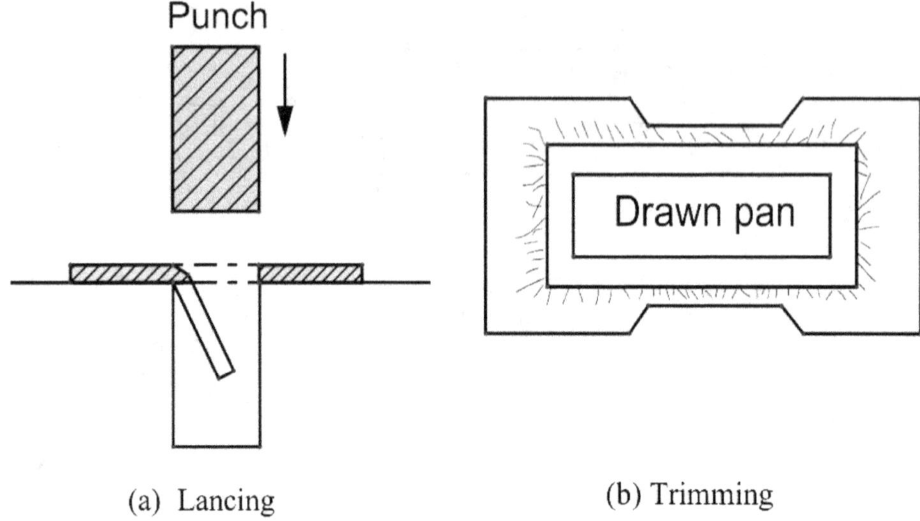

(a) Lancing   (b) Trimming

Figure 10.6 Lancing and trimming operations

10.4.9 Trimming

During sheet metal working operations, such as blanking and punching, the sheared sheet usually bears uneven edges, which are also known as burrs. These burrs need to be removed to make the sheet metal part useful. The trimming operation is used for the removal of burrs attached to the edges of the workpiece after the sheet metal working operations is over. The trimming operation therefore is also known as the shaving operation as it makes the edges to be clean and smooth free from burrs. Figure 10.6 (b) shows a snapshot of the trimming operation.

10.4.10 Riveting

Riveting operation (Figure 10.7) is used to join sheets with the help of rivets. This operation can either be done manually or with the help of a riveting gun.

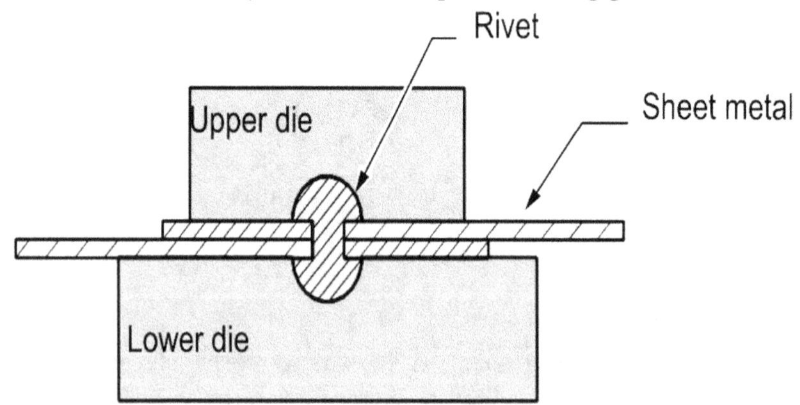

Figure 10.7 Riveting of sheet metal plates

10.5 SAFETY PRECAUTIONS FOR MACHINING

The precautions to be followed in sheet metal shop are mentioned below.

i. Most of the sheet metal working operations make use of shearing and sharp-edged tools. Extra care is therefore needed to keep hands and fingers safe while working in this sheet metal shop.

ii. For marking purposes, use scriber or trammel only. Do not use a pencil or a pen.

iii. Sufficient care is to be taken while cutting and folding of galvanized iron (G.I.) sheet.

iv. Immediately remove the scrap and waste pieces from the workplace.

v. Make use of hand leather gloves while handling heavy metal sheets.

vi. Avoid pulling of the sheared sheet part by hand while cutting with snip.

vii. Be cautious and avoid inhalation of toxic cadmium fumes which are present in greater quantity when plated materials are brazed.

PRACTICE NO. 10.1

Job: To make a lap joint of two metal sheets using riveting

Objectives: Learn by practice making of the sheet metal lap joint using the riveting operation.

Machines, equipment, and tools required: Try-square, scriber, straight snip, 300 mm steel rule, rivet hammer and rivet.

Material required: Galvanized iron (G.I.) sheet of 100 mm x 80 mm size.

Schematic of the job:

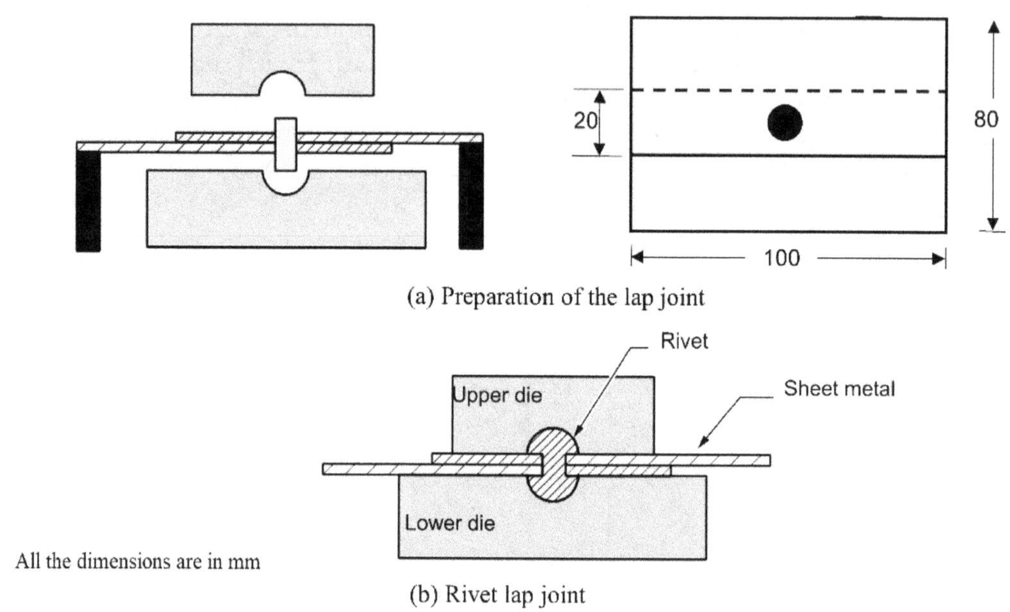

Figure P10.1 Lap joint of two metal sheets using riveting

Procedure:

i.   Take a thin 24-gauge G.I. sheet of 100 mm x 80 mm size. Cut the sheet with the help of a hand shear or bench shear if the required size is not readily available.

ii.   Mark the sheet with the help of scriber and punch a hole in both the sheets.

iii.   Place the sheets and the rivet as per the description shown in Figure P10.1 (a) in between the lower and the upper halves.

iv.   Strike the upper die using a riveting hammer till the tail is deformed and a head is created.

v.   Remove the upper die and check the rivet heads.

vi.   Fill out the workshop practice response sheet given at the end of Chapter 2 and answer the questions provided therein. Get feedback of the instructor on the job prepared by you.

PRACTICE NO. 10.2

Job: To make a sheet metal tray of given dimensions using sheet metal working operations.
Objectives: Learn by practice, the sheet metal working operation for making a tray of given dimensions.
Machines, equipment, and tools required: Divider, try-square, scriber, straight snip, steel rule and mallet.
Material required: G.I iron sheet of 100 mm x 100 mm size.
Schematic of the job:

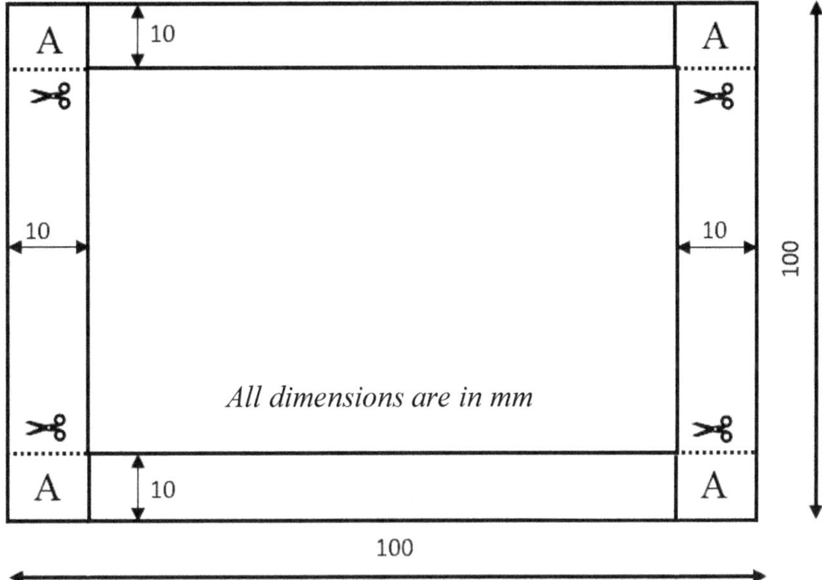

Figure P10.2 (a) Dimensions of sheet metal tray workpiece

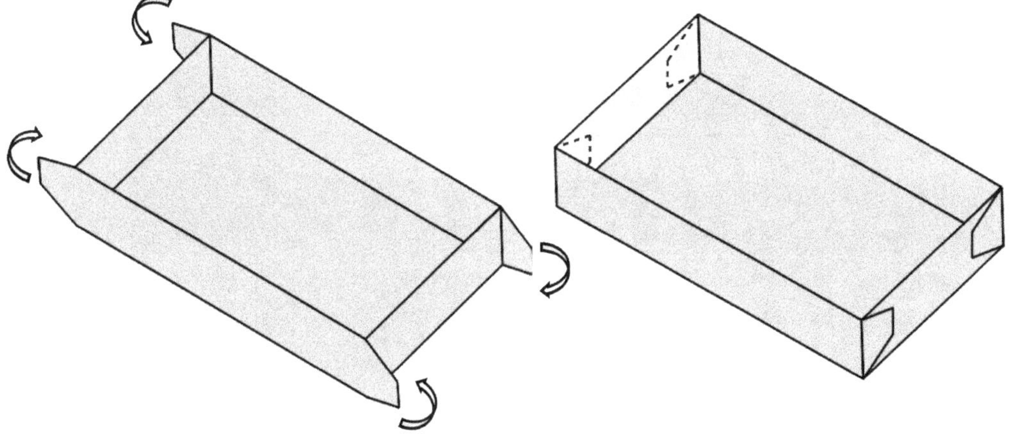

Figure P10.2 (b) The workpiece after shearing and folding

Procedure:

i. Take a thin 24-gauge G.I. sheet of 100 mm x 100 mm size. Cut the sheet with the help of a hand shear or a bench shear if the required size is not readily available.

ii. Mark the sheet with the help of a scriber in such a manner so as to leave a margin of 10 mm on both left and right side, and 10 mm each on both the top and bottom sides. The markings are shown in figure P10.2.

iii. Make two horizontal cuts of 10 mm length each in (marked as A) starting from left and right sides along the marked lines. A total of four cuts are to be made with the help of a snip (see Figure P10.2 (b)).

iv. Bend the sheet area from the top and bottom marking of 10 mm margin with the help of a hammer (or a mallet) by keeping the workpiece on the anvil (or stake).

v. Bend the sheet area from the left and right marking of 10 mm margin with the help of a hammer (or mallet) and by keeping the workpiece on the anvil (or stake).

vi. The corner portions of the sheet are folded to come inside and align with the vertical walls so formed by bending the 10 mm margin.

vii. The workpiece now looks like a tray. Now, fold the top 10 mm of the left and right margin inside the tray (see Figure 10.2 (b)).

viii. Riveting of the sheets is done to make it strong. The procedure given in practice 10.1 can be used for riveting.

ix. Alternatively, soldering may be done if a leak proof joint is needed. The procedure for soldering is already discussed in practice 7.2.

x. Fill out the workshop practice response sheet given at the end of Chapter 2 and answer the questions provided therein. Get feedback of the instructor on the job prepared by you.

PRACTICE NO. 10.3

Job: To prepare a funnel with sheet metal working and soldering operations
Objectives: Learn by practice various tools and operations required for making a funnel.
Machines, equipment, and tools required: Steel foot rule, spring divider, straight snip, mallet, chisel hammer, dot punch, half round file, bench vice, anvil, lever shearing machine, soldering iron, funnel stake, plier, nose plier and long nose hammer.
Material required: Galvanized iron sheet of SWG 28 (0.32 mm)
Schematic of the job:

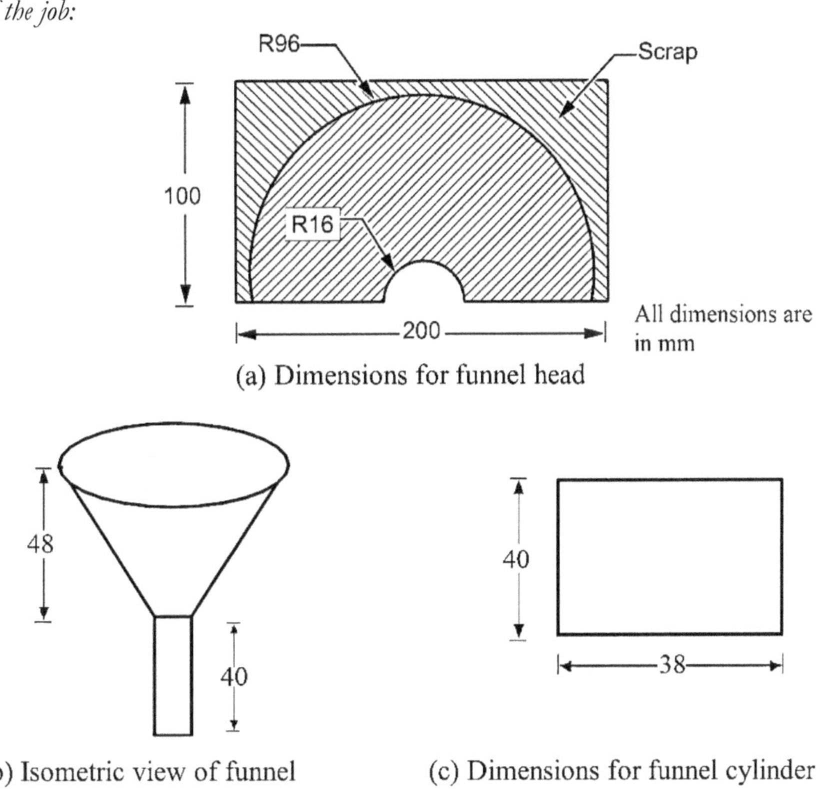

Figure P10.3 Funnel made with sheet metal working

Procedure

    i.   Observe the schematic provided in Figure P10.3 and get a GI sheet workpiece of size 200 mm x 100 mm taking help of a lever shearing machine.

    ii.   Draw a semi-circle radius of 16 mm radius and another of radius 96 mm as shown in the Figure P10.3 (a)

    iii.   Punch outline of both the circles with the help of a dot punch and hammer.

    iv.   Cut along both the circles with the help of a straight snip tool.

    v.   Now mark a 5 mm line on two straight edges of semi workpiece and fold them in the opposite direction.

    vi.   With help of a mallet and stake (or anvil) bend the sheared workpiece to make a conical shape and bring both the folded ends near to each other.

    vii.   Now insert the folded ends into each other to make a seam joint.

    viii. Bend the sheet on the taper stake so that the locking of the joint is done.

ix.   Cut another G.I. sheet workpiece of size 40 mm x 38 mm and bend it along its length to form a cylindrical shape slightly folding its ends to make them join.

x.   Keep one end of the cylindrical workpiece in the hole of the funnel and solder the joint.

xi.   Fill out the workshop practice response sheet given at the end of this chapter 2 and answer the questions provided therein. Get feedback of the instructor on the job prepared by you.

Chapter 11

POWER TOOLS

11.1 INTRODUCTION

Power tools are very commonly used in a variety of engineering applications. These tools are being supplied by several manufacturers and are available at home improvement or hardware stores. The selection of a power tool needs to be done very carefully keeping in view the intended application. Learning of these tools require the user to follow strict safety precautions and use them under the supervision of a trained technical staff. Utmost care is required to keep one's fingers, hands, and body at a safe distance from any moving parts.

The commonly used power tools are discussed in the following paragraphs.

11.2 POWER DRILL

Power drill is the most commonly used tool in many applications in home improvement, garage work and workshop operations. Figure 11.1 (a) shows a snapshot of a power drill, which may be either of cordless type or with cord. Most of the work done with this tool is for making holes and driving screws in concrete and wood. The power drill can also be used for applications like mixing of paints.

11.3 ROTARY HAMMER

Rotary hammer is used for carrying out heavy-duty jobs like drilling and chiseling for making holes for cutting action. A rotary hammer will provide a lot of impact energy than the power drill. Figure 11.1 (b) shows a snapshot of a rotary hammer.

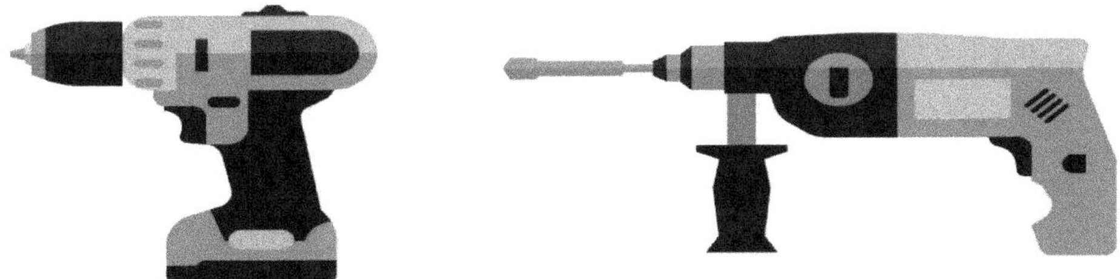

(a) Power drill   (b) Rotary hammer

Figure 11.1 Snapshots of power drill and rotary hammer

11.4 IMPACT WRENCH

Impact wrench is also known as an impact gun as it provides high torque output using minimal physical action. This is done by storing the energy in a rotational mass followed by its quick release to the output shaft. The impact wrench finds applications in several industries where high torque is required, such as the assembly of automobiles, heavy equipment maintenance and major construction projects. Many impact wrenches also come with the provision of adjustment of the output torque to prevent damage of parts they are going to fasten. Figure 11.2 (a) shows a snapshot of an impact wrench.

11.5 JIGSAW

Jigsaw is a compact power tool which finds its applications in making cuts along a curve in wood and other materials. Although most saws are supposed to make linear cuts, the jigsaw allows to cut shapes which are more complicated and delicate. Figure 11.2 (b) shows a snapshot of a jigsaw.

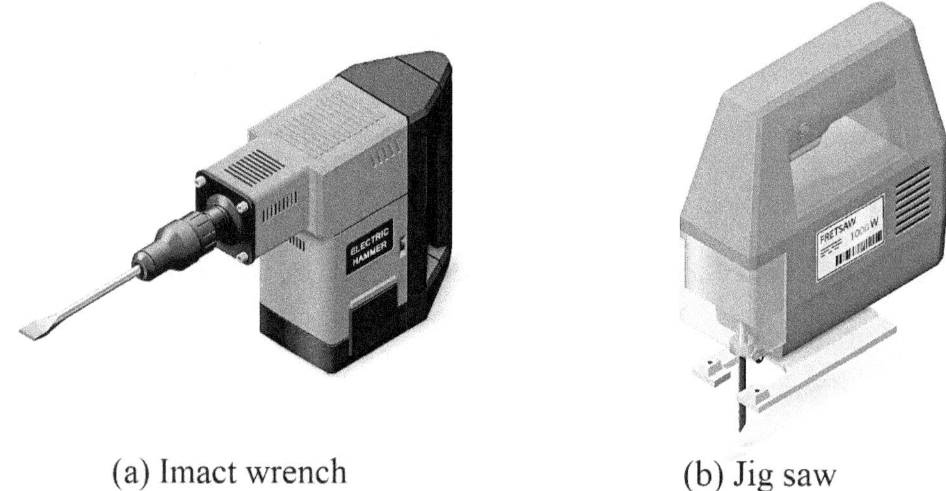

(a) Imact wrench   (b) Jig saw

Figure 11.2 A snapshot of impact wrench and jig saw

11.6 ELECTRIC SCREWDRIVER

Electric screwdriver makes use of a motor to drive the screwdriver. This is generally a compact device which also has a rechargeable battery allowing its cordless function. Figure 11.3 (a) shows a snapshot of an electric screwdriver.

11.7 BAND SAW

Band saw is a power tool that has a fixed structure and a sharp long blade which provides a continuous cutting action. The material to be cut is supported on a table provided with the band saw. The cutting blade of this saw is made of a flexible yet hard steel which provides cutting action by rotating. A common application of this saw is wood working, but it can also be used for cutting other materials. Figure 11.3 (b) shows a snapshot of a band saw.

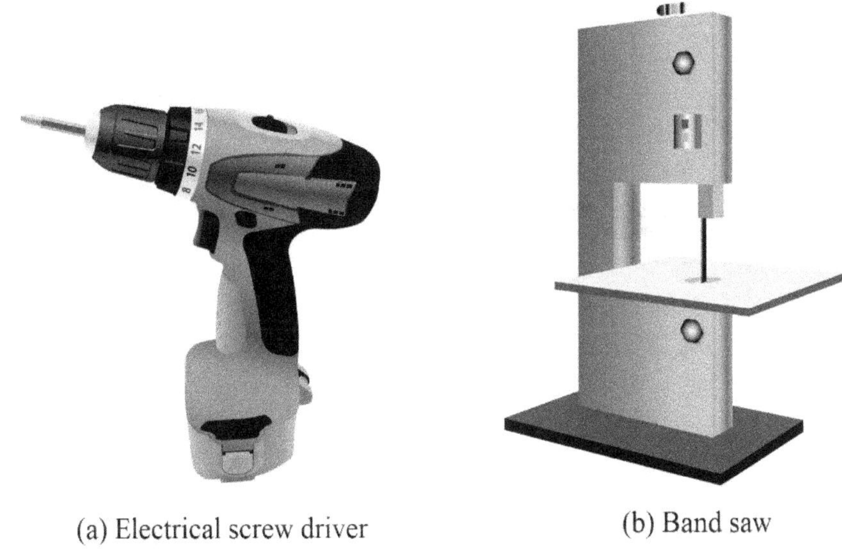

(a) Electrical screw driver   (b) Band saw

Figure 11.3 Snapshots of electric screwdriver and band saw

11.8 TABLE SAW

A table saw also known as a bench saw is commonly used for woodworking. It has a circular saw blade mounted on its arbor. An electrical motor is used to drive the saw in a circular fashion to cut material. Only a small portion of the saw, which can be adjusted though, comes out of the tabletop for performing the cutting action. Figure 11.4 (a) shows a snapshot of a table saw.

11.9 CIRCULAR SAW

A circular saw uses a rotary toothed blade which is in the form of a disc fixed to an arbor for making different cuts in a material with a rotary motion that spins around an arbor. The most common application of this tool is to cut metals, plastic, masonry, and wood. Figure 11.4 (b) shows a snapshot of a circular saw.

11.10 HEAT GUN

Heat gun is used to beam hot air at a particular target just like a hairdryer but the air in this is very hot. It is commonly used to bend plastic and heat frozen pipes. Figure 11.4 (c) shows a snapshot of a heat gun.

11.11 ANGLE GRINDER

Angle grinder is a portable device also known as a side grinder, which is used to grind and polish material. Applications of this grinder are very wide as it can be used for jobs which cannot be brought to the bench grinder. Figure 11.4 (d) shows a snapshot of an angle grinder.

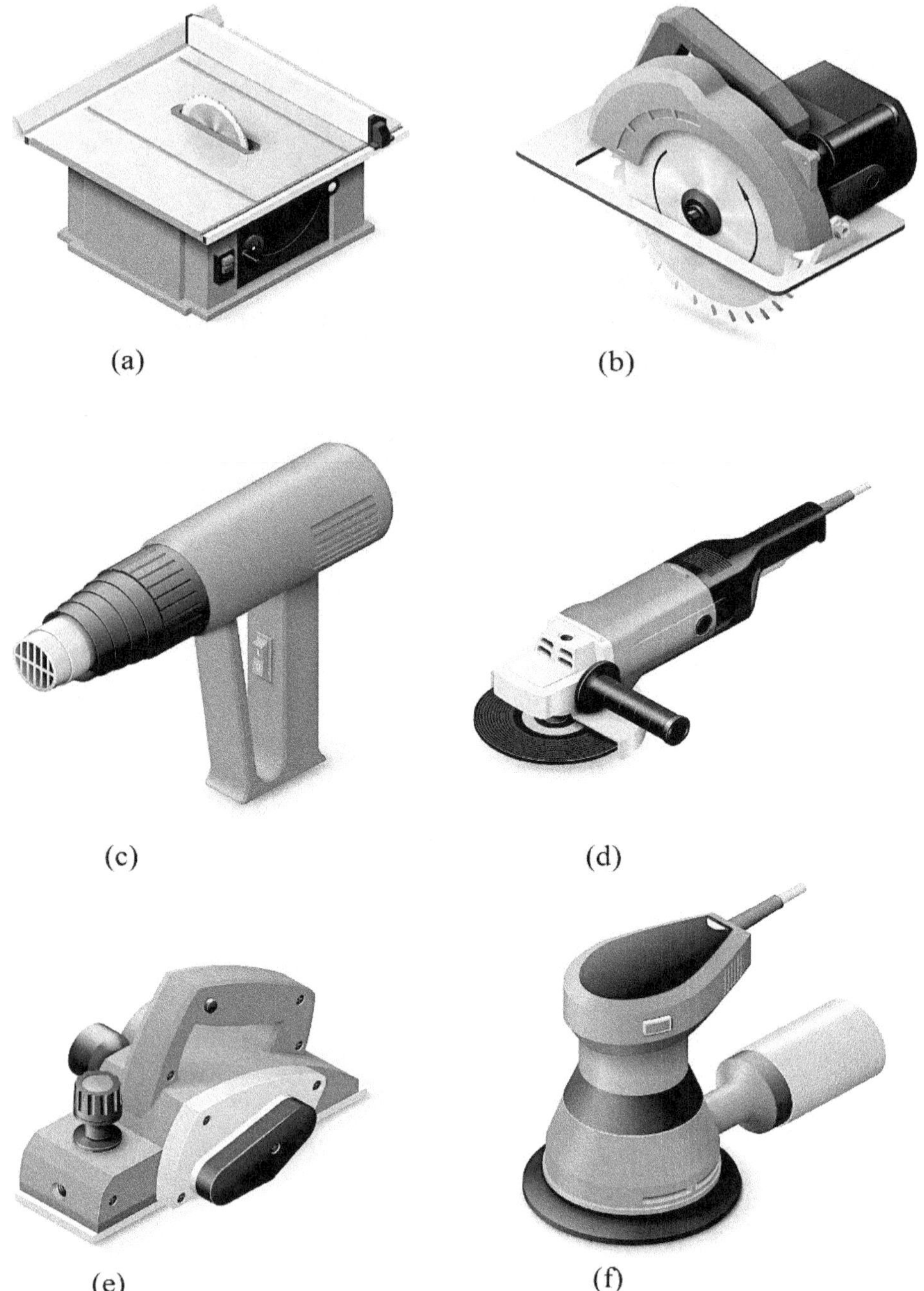

Figure 11.4 Hand power tools (a) table saw, (b) circular saw, (c) heat gun, (d) angle grinder, (e) power planer, (f) disc sander

11.12 POWER PLANER

Power planer (Figure 11.4 (e)) is used for planning of wooden jobs and is a replacement of the jack plane that was discussed in Chapter 2.

11.13 DISC SANDER

Disc sander is a power tool which replaces the manual sanding action. This tool has a round piece of replaceable sandpaper which is attached to a circular wheel. The sanding action is performed on a job with the help of rotational motion of the wheel and the sandpaper. Figure 11.4 (f) shows a snapshot of a disc sander.

11.14 CHAIN SAW

As the name appears, a chain saw is a portable saw which has a set of teeth which are attached to a rotating chain. The rotating chain moves along a guide bar to perform cutting action. The most general use of chain saw is cutting trees and wood logs. Figure 11.5 (a) shows a snapshot of a chain saw.

11.15 RECIPROCATING SAW

Reciprocating saw performs cutting action with the help of a mechanism that applies the push and pull concept. Its most common use is in construction work and demolition of structures. Figure 11.5 (b) shows a snapshot of a reciprocating saw.

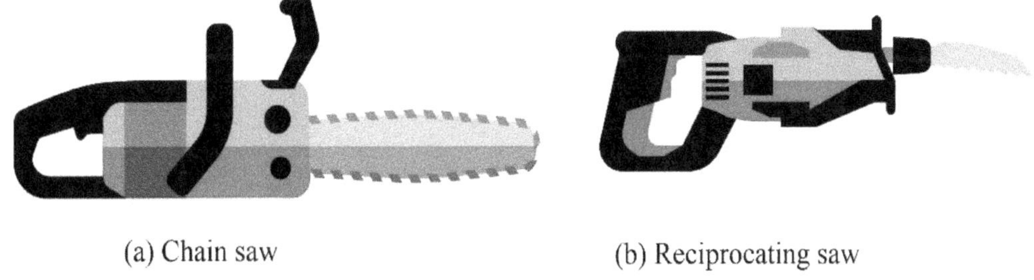

(a) Chain saw   (b) Reciprocating saw

Figure 11.5 Snapshots of chain saw and reciprocating saw

11.16 BENCH GRINDER

Bench grinder is used for grinding mostly metallic jobs with the help of grinding wheels which are provided with motorized rotation. The grinding wheels has abrasive materials which are securely bound on a metallic wheel. The most common application of a bench grinder is the removal of the extra layer of material on a job by grinding action thereby converting the material to small dusty particles. The bench grinder is commonly used to sharpen hard cutting tools too. Figure 11.6 (a) shows a snapshot of a bench grinder.

11.17 NAIL GUN

Nail gun is a power tool which is used for inserting nails into an object with forceful action. It is a replacement for a hammer and is useful for projects that require nails to be applied. Figure 11.6 (b) shows a snapshot of a nail gun.

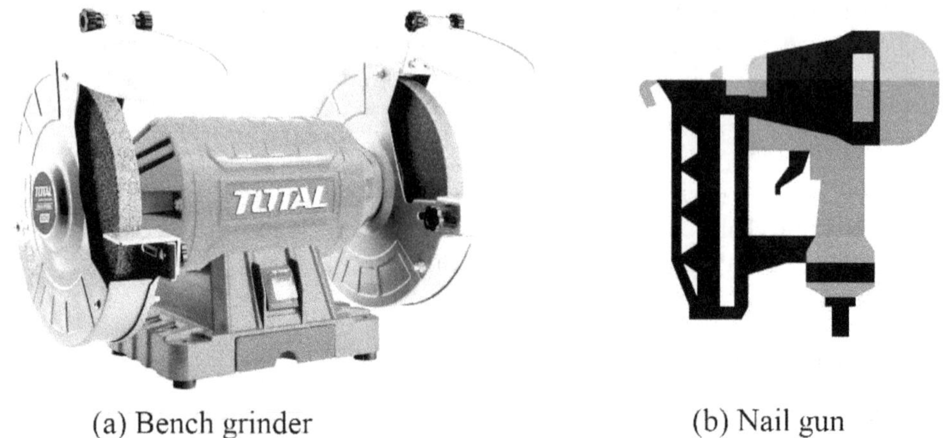

(a) Bench grinder (b) Nail gun

Figure 11.6 Snaphsots of bench grinder* and nail gun

(*Courtesy: Total Tools, Kochi, Kerala, India, used with permission
weblink: TOTAL (totalpowertools.in))

11.18 AIR BLOWER

Air blowers are very commonly used for blowing air to perform certain tasks, such as blowing off horticultural waste to move them so that it can be collected or pushed to a collecting device. It is also used for applications like cooling, drying, pollution control, and many such tasks. In carpentry shop, an air blower is used to clean wood scrap from the table. Figure 11.7 shows a snapshot of an air blower.

Figure 11.7 Snapshots of air blower

(Courtesy: Polymak Tools India Pvt. Ltd., Chennai, India, used with permission.
Weblink: https://www.polymak.co.in/product/pm600br/)

QUESTIONS

Q11.1 Name the power tools which can be used for creating holes in a structure.

Q11.2 Which tool can be used for inserting nails into an object?

Q11.3 Name the tool which is used for tightening of screws and nuts.

Q11.4 Name the tools which are used for blowing air, whether hot or at normal temperature.

Q11.5 Which tools are used for wood working tasks? Which function or tasks they can do?

PRACTICE NO. 11.1

Student's name: Roll Number:

Objective: Understand the functioning and control of all the power tools described in this chapter and use them for their intended application.

Power Tools Response Sheet
Write down the names, control features and applications of the power tools you used.

Instructor remarks

Date: Signed/Checked by

Chapter 12

3D PRINTING

12.1 INTRODUCTION

3D printing is a relatively new manufacturing process which provides ample freedom to the designer for designing a part by selecting complex features which may not be manufacturable using conventional manufacturing processes. In this process, a part is built layer by layer. Material is deposited in the lowermost layer first followed by the next layers. This continues until material at the top layer is deposited, thus completing the part. A snapshot of the 3D printing process is shown in the following Figure 12.1.

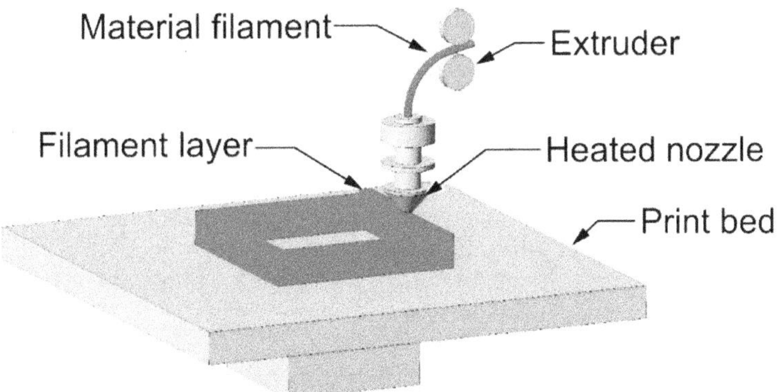

Figure 12.1 A schematic to showcase 3D printing process

12.2 TECHNOLOGIES FOR 3D PRINTING

This section discusses different 3D printing technologies. A common concept followed in these technologies is that they create parts by essentially depositing material layer by layer, although they may utilize a different type of material resulting in varying properties. Some of the most common and utilized types of technologies are Fused Deposition Modelling (FDM), Selective Laser Sintering (SLS), Selective Laser Melting (SLM), Stereolithography and Digital Light Processing (DLP). The following sections provide a brief overview of the important 3D printing technologies.

12.2.1 Fused deposition modelling (FDM)

In this type of process, the plastic material which is in the form of a filament is deposited and brought to the plastic stage by passing it through a hot metal nozzle (also called the extruder). The material is deposited layer by layer by controlling the movement of the nozzle and the platform until desired shape and size of the part are obtained. There are several materials which can be used in the FDM process; however, ABS and PLA are the most common.

12.2.2 Selective laser sintering (SLS)

Selective laser sintering uses a laser beam to sinter powdered plastic material and turns it into a solid model. Normally, this type of technology is a popular choice for making prototypes and small-batch manufacturing. The advantage of this process is that the un-sintered powder can be reused, and higher accuracy of the parts produced.

12.2.3 Selective laser melting (SLM)

Selective laser melting uses a high-power density laser to melt and fuse metallic powders to obtain joining. SLM completely melts the materials to produce a solid metallic part. The advantage of the SLM process is that metallic parts can easily be made, which is otherwise done by the casting process.

12.2.4 Stereolithography (SLA)

Stereolithography is used to create parts with high levels of detail, smooth surfaces, flawless finishes, and quality. This process uses liquid in place of powder, whereas laser power is used to solidify the material layer by layer.

12.2.5 Digital Light Processing (DLP)

Digital light processing is a technique like SLA that cures the resin materials with the help of light through a light projector screen. Because of the light usage, an entire layer can be built at once making this process relatively faster. However, this process is recommendable mostly for low-volume production runs of plastic parts.

12.3 PARTS OF A 3D PRINTER

A 3D printer has several components, which depend on the technology being used in it. However, all 3D printers need mechanisms for three-degree freedom to print a three-dimensional part. Since the scope of this chapter is limited to introducing the 3D printing technology, we describe a 3D printer, which follows FDM process explained in the previous section. For an FDM printer, the most important components are the printing bed (or platform), extruder (or nozzle), moving parts for three-dimensional axis control, and graphic user interface (GUI) panel for interaction with the user. Details of important parts of a 3D printer are below mentioned.

    i.   *Printing bed*: The printing bed is the platform for printing models. In an FDM printer, the printer bed is heated, which helps in the adhesion of plastic material layers to the lowermost layer of the platform as well as amongst themselves.

    ii.   *Extruder*: The extruder is the core component of a 3D printer, which will melt and stretch the filaments to build the model.

iii.   *Moving parts*: The parts of the printer will move on three axes, which are X, Y, and Z axis. The X and Y axes represent forwarding and backward movements associated with the extruder. The Z-axis represents the vertical movements required for controlling the platform.

iv.   *GUI panel*: Users can operate the printer and complete various settings by clicking the built-in interface provided on the LED touch screen.

Figure 12.2: Snapshot of a 3D printer and its important parts

(Courtesy: Raise 3D Technologies, Inc., Printer model - RAISE 3D Pro3, used with permission, weblink: https://www.raise3d.com/products/pro3/)

12.4 PROCEDURE OF 3D PRINTING

In short, the 3D printing process required four steps, as mentioned below.

   i.   Preparing CAD model of the part

  ii.   Converting CAD file to the STL file format

 iii.   Slicing and tool path generation

 iv.   3D printing of the part.

The above-mentioned steps have been explained with the help of the following paragraphs with illustrations.

Step I - Preparing CAD model of the part: This step involves preparing CAD model of the desired part in a CAD modeling software. Examples of CAD modeling software are AutoCad, FreeCAD, SolidWorks, CATIA, SolidEdge, Creo and NX. Several other CAD modeling software are available which may also be used to prepare a part's CAD model. Figure 12.3 (a) shows a snapshot of a CAD model.

Step II - Converting CAD model to the STL file format: In this step, the part's CAD model file is converted to the STL format (also called stereolithography file format). The STL file format is acceptable to the slicing software. It is to be noted that '*save as*' or '*export*' options available in FILE menu of the software may be used for converting CAD model in the STL file format. In STL file format, each surface of the CAD model is converted into triangles. You may be aware that a triangle is composed of lines in a plane, which is helpful in generating slicing. Figure 12.3 (b) shows a snapshot of the STL model of the example part.

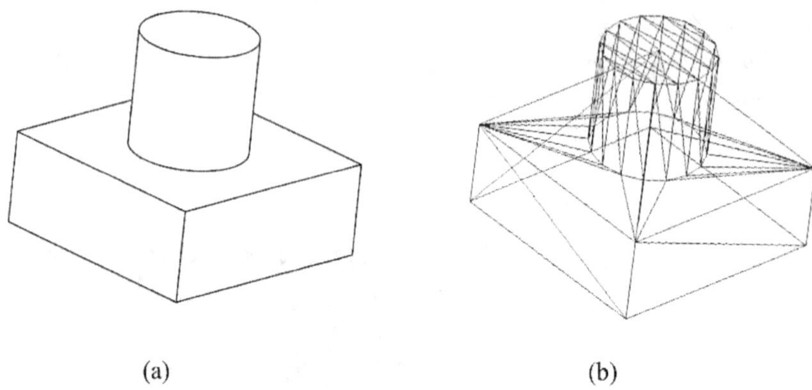

(a)   (b)

Figure 12.3 Snapshot of (a) CAD model of an example part (b) it's STL file

Step III - Slicing and tool path generation: In this step, the STL file of the part is imported to the slicing software. Several slicing software are available, many of which are free of cost. If you don't have a slicing software installed on your computer, you can easily search for one using a web search engine, download, and install it on your computer. Examples of slicing software are Cura, 3DPrinterOS, and IdeaMaker.

During the installation process, you may be asked a few questions, such as *'select your printer'* and *'select the filament'*. You may answer the questions yourself in case you know else contact your instructor in this regard

The STL file is imported to the slicing software to obtain the slicing of the part model. This step typically generates several slices of the part model. Each slice of the part is helpful to get the requisite information for depositing the material for printing the part slice by slice. Figure 12.4 shows a snapshot of the slicing of the part model done in the slicing software.

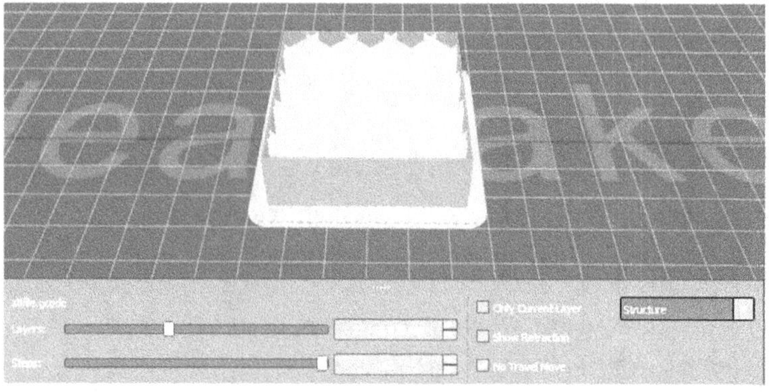

12.4: Snapshot of the slicing of the part model

Thereafter, the slicing file is exported as a *'gcode'* file to be saved on the computer to transfer to the 3D printer using a USB drive or another communication mechanism. Figure 12.5 shows a representative snapshot of the *'gcode'* file (the actual file is very long), which has statements to direct the extruder to go to the specified location for depositing the material for 3D printing.

```
; Sliced by ideaMaker 4.3.0.6190, 2022-08-22 22:15:19 UTC+0530;; Origin Center: 0

; Extruder Offset #1: 25.000 0.000
; Filament Diameter #1: 1.750
---------------------------------
M221 T0 S94.00
M140 S60.00

---------------------------------
M109 T0 S215.00
T0
M190 S60.00
G21
G90

---------------------------------
```

Figure 12.5 A representative snapshot of the '*gcode*' file for 3D printing

Step IV- 3D printing of the part: This is the last step in which the '*gcode*' file generated in the previous step is used for printing the part on the 3D printer for which the following steps are used.

    i.   Switch on the 3D printer
    ii.   Place your USB memory stick containing the 'gcode' file in the USB drive. You may use another method of data transfer in place of USB memory stick, such as a wireless connection.
    iii.   Clean the surface bed plate (or platform) of the 3D printer
    iv.   Load the required filament into the extruder.
    v.   Wait until the nozzle and bed plate comes to the required temperature.
    vi.   Close the doors of the 3D printer.
    vii.   Open the file from the USB storage and select the 'print' option
    viii.   After the part is printed, remove it from the platform using the given tools
    ix.   Clean the part and remove extra materials attached to the part.

Note: You may be required to select 3D printing information, such as infill density and other steps for printing the part on 3D printer. It is recommended to refer to the 3D printer manual for the same.

PRACTICE NO. 12.1

Job: To prepare a CAD model of a simple part with your name and roll number embossed/engraved and print it on a 3D printer.

Objectives: To learn by practice the procedure to make a simple CAD model and its 3D printing

Machines, equipment, and tools required: FDM printer, filament PLA/ABS, CAD modelling software (SolidWorks/Inventor/FreeCAD), Slicing software (Cura/Ideamaker), Computer.

Material required: PLA 1.75 mm.

Schematic of the job:

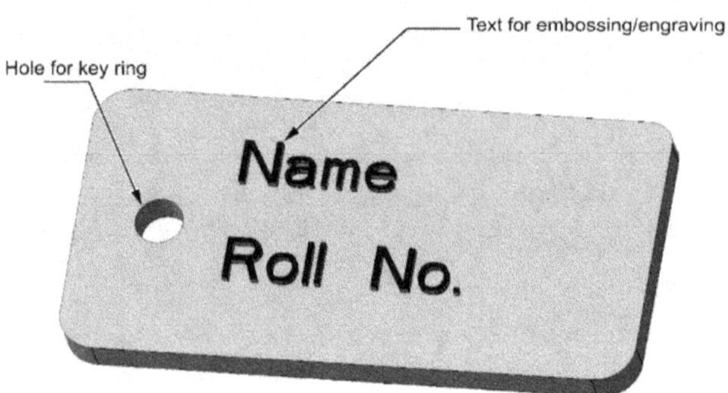

Figure P12.1 CAD model of a simple part for 3D printing

(Important note: You may select a different part design for 3D printing with the approval of the instructor. Innovation and application of your artistic and engineering skills is encouraged*)*

Procedure:

i.   Install a CAD modelling software on your PC/Laptop/Mac/Computer (AutoCad/Inventor/FreeCAD/Solidworks/Creo)

ii.   Prepare CAD model of the part to be printed with the with approximate size of 60 mm x 30 mm x 4 mm size with corners filleted with 2 mm radius. Also place a hole of about 4 mm diameter for inserting keyring into it. Your name and roll number should come on the top of the block with a thickness of 3 mm (either embossed or engraved). Other dimensions which are not mentioned may be suitably assumed. Figure P12.1 shows snapshot of the part to be made.

iii.   Save or export the part CAD model in the STL file format.

iv.   Install a slicing software, such as Cura or IdeaMaker.

v.   Import the STL file of the part model prepared in step iii into the slicing software. Perform slicing of the file by selecting suitable parameters, such as layer thickness.

vi.   Generate toolpath (gcode file) for printing the part on the 3D printing machine

vii.   Use the toolpath file obtained in step vi to print the part on a 3D printing machine. FDM printer is preferred as it is very simple and easily available.

viii.   Reclaim the 3D printed part from the machine, clean it and remove extra materials such as the support base attached to it (also called as raft).

ix.   Fill in the 3D printing practice response sheet given at the end of this chapter and answer the questions provided therein. Get feedback of the instructor on the job prepared by you.

3D PRINTING RESPONSE SHEET

Student's name: Roll Number:

Job Name: Job material:

Objective of this practice:

Describe the stepwise procedure you adopted for 3D printing of the part? Also state the software tools you used in each step.

What did you learn in this practice sessions?

State the safety precautions followed in this practice session.

Carefully inspect the job and note the shortcomings/defects found in consultation with the instructor.

Instructor remarks

Date: Signed/Checked by

MULTIPLE CHOICE QUESTIONS
(Select one correct option)

1. Identify the type of tong as shown in the figure.

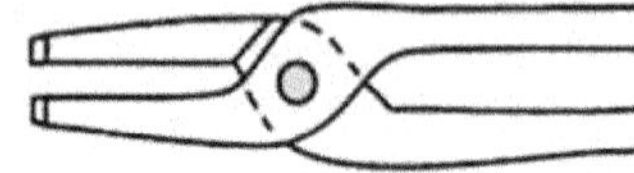

   (a) Open mouth tong
   (b) Closed mouth tong
   (c) Square hollow tong
   (d) Round hollow tong

2. Which software is used for slicing during 3D printing of parts?
   (a) Curo
   (b) IdeaMaker
   (c) 3D printer OS
   (d) All of the above

3. What is the purpose of straight peen hammer?
   (a) Heavy duty work
   (b) Stretching the workpiece
   (c) Riveting and chipping
   (d) Bending

4. Identify the electric switch shown in the figure.

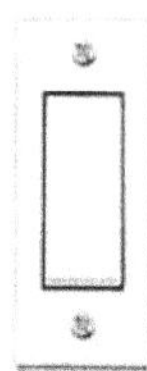

   (a) One way switch
   (b) Two way switch
   (c) Power switch
   (d) Bell switch

5. What is the primary function of anvil used in the smithy shop?
   (a) Cool the object to the desired temperature
   (b) Allow the passage for flue gases
   (c) Allow the air supply into the furnace
   (d) Provide support to the job to work on it in hot conditions

6. What is the primary function of anvil used in the smithy shop?

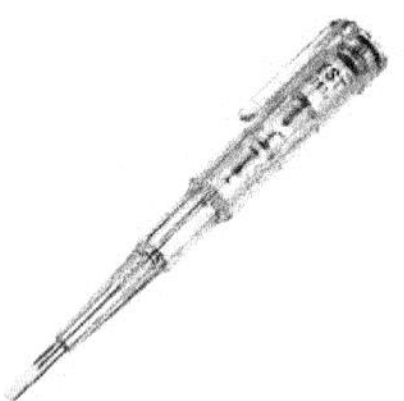

   (a) Insulated tape
   (b) Voltage tester
   (c) Wire stripper
   (d) Soldering iron

7. Which of the following is/are the types of power tools?
   (a) Power planner
   (b) Rotary hammer
   (c) Power drill
   (d) All of the above

8. Angle grinder tool is used for_____________?
   (a) Grinding and polishing the material
   (b) Cutting of wooden jobs
   (c) Striking the wooden jobs
   (d) Planning the wooden jobs

9. Which of the following is the material used for a FDM 3D printer?
   (a) Extruder
   (b) Printing bed
   (c) Glue
   (d) PLA filament

10. Which of the following is/are the parts of a 3D printer?

(a) Printing bed
(b) Extruder
(c) GUI panel
(d) All of the above

11. Which of the following is/are the different types of 3D printing technology?
    (a) Fused deposition modelling
    (b) Selective laser sintering
    (c) Stereolithography
    (d) All of the above

12. Which of the following is/are the cutting parameters for machining on a lathe machine?
    (a) Cutting speed
    (b) Feed
    (c) Depth of cut
    (d) All of the above

13. What is the function of lathe bed in a lathe machine?
    (a) Support the main spindle
    (b) Hold the cutting tools such as drills and chuck
    (c) Control the movement of the cutting tool in transverse and longitudinal direction
    (d) Provide rigid support to the parts of lathe machine

14. Identify the power tool as shown in figure

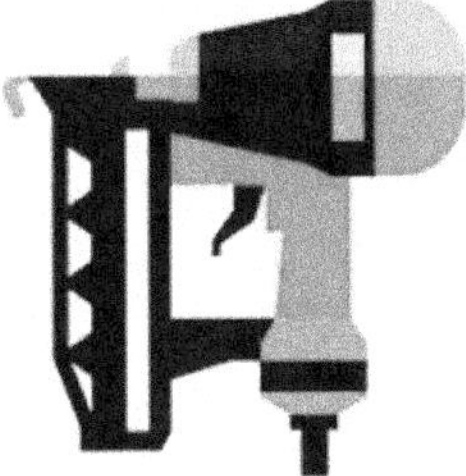

(a) Chain saw
(b) Hacksaw
(c) Heat gun

(d) Nail gun

15. Identify the electrical symbol used in drawing the electrical circuit.

(a) Cell
(b) Switch
(c) Bulb
(d) Fuse

16. Among the following operations, which operation is not done on a swage block.
    (a) Sizing
    (b) Squaring
    (c) Bending
    (d) Welding

17. What is the function of compound slide in a lathe machine?
    (a) Provide rigid support to the parts of the lathe machine
    (b) Hold the cutting tools such as chuck and drill.
    (c) Control the movement of the cutting tool in transverse and longitudinal direction
    (d) Support the tool post and its movement in the angular direction to get tapered surfaces.

18. Molten metal is poured into a cavity known as
    (a) Riser
    (b) Runner
    (c) Casting
    (d) Mold

19. Which among the given operation is not a type of machining operation
    (a) Soldering
    (b) Grooving

(c) Drilling
(d) Taper turning

20. The tools used in carpentry shop is/are
   (a) Marking gauge
   (b) Try square
   (c) Mortise gauge
   (d) All of the above

21. For the effective cutting, the cutting blade of a wooden jack should be inclined to
   (a) 45 degrees
   (b) 30 degrees
   (c) 60 degrees
   (d) 90 degrees

22. Which of the following is not a type of striking tool?
   (a) Claw hammer
   (b) Mallet hammer
   (c) Cross peen hammer
   (d) Jack plane

23. The baldes of the chisel are generally made from
   (a) Forged steel
   (b) Nickel
   (c) Aluminium
   (d) Mild steel

24. Identify the component 'A' shown in the given figure

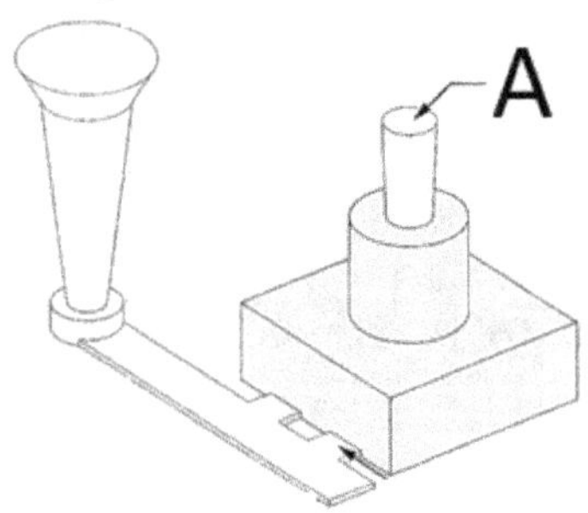

   (a) Riser
   (b) Runner

(c) Gate
(d) Poring basin

25. Identify the welding defect shown in the figure below

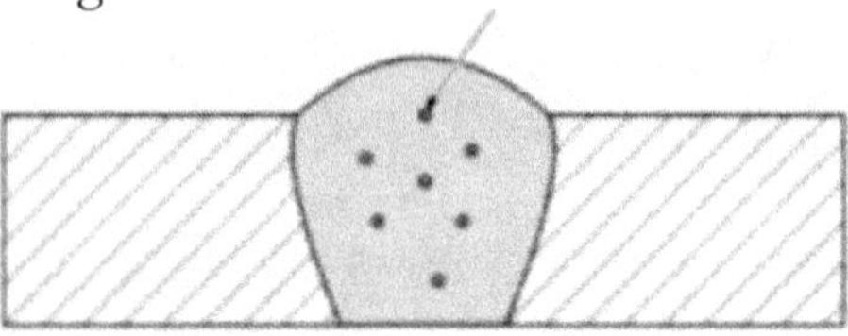

   (a) Cracks
   (b) Porosity
   (c) Inclusion
   (d) Undercut

26. Which of the followings is not a work holding device?
   (a) Bench vice
   (b) Pipe vice
   (c) Scriber
   (d) Workbench

27. Which among the below is not a type of marking tool in fitting shop?
   (a) Steel rule
   (b) Divider
   (c) Bench vice
   (d) Try square

28. Identify the type of file shown in the figure below

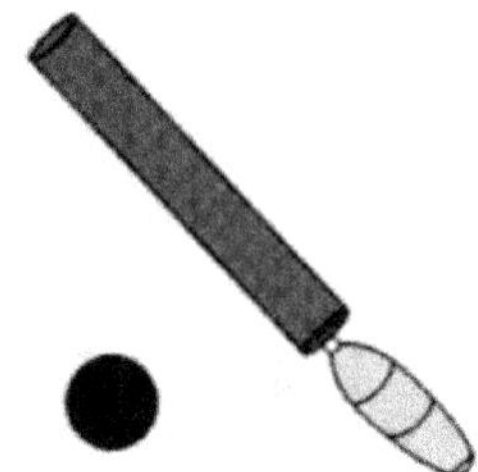

   (a) Flat file
   (b) Triangular file
   (c) Square file
   (d) Round file

29. Which of the following is not a sheet metal operation?

(a) Embossing
(b) Forging
(c) Blanking
(d) Punching

30. Identify the type of hammer as shown in figure used for sheet metal operations

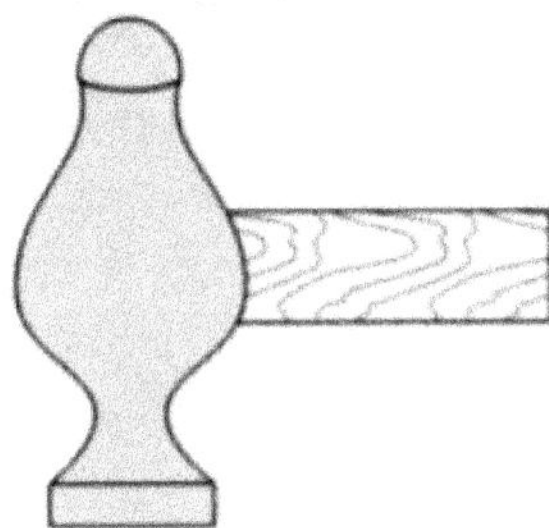

(a) Square head hammer
(b) Cross peen hammer
(c) Straight hammer
(d) Ball peen hammer

31. Identify the warning symbol as shown in figure

(a) Nose mask
(b) Emergency exit
(c) Face shield
(d) Danger

32. For the protection of eyes, you need to wear?
(a) Safety shoes
(b) Hand gloves
(c) Safety glasses
(d) Nose mask

33. Identify the warning symbol in the figure below

(a) Flammables
(b) Health hazard
(c) Emergency exit
(d) Danger

34. The concept used in machining screw threads on lathe machine is
(a) The tool is moved by operator manually
(b) The tool moves with the help of lead screw and nut arrangement
(c) Tail stock helps in motion of the tool
(d) The tool moves like normal plane turning operation

35. Match the following tools as per application

(a) Jack plane	(a) Cutting tool
(b) Mallet	(b) Planning tool
(c) Carpenter's vice	(c) Striking tool
(d) Saw	(d) Work holding device

36. Identify the electrical symbol used in drawing the electrical circuit.

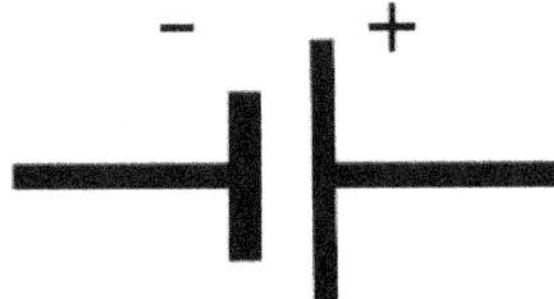

(a) Switch
(b) Fuse

(c) Bulb

(d) Cell

37. Which of the following is not a power tool?

    (a) Hacksaw

    (b) Power drill

    (c) Power planner

    (d) Rotary hammer

38. Which file type is most commonly used as input for 3D printing?

    (a) TIFF

    (b) JPG

    (c) X3G

    (d) STL

39. Which of the following are CAD modelling software?

    (a) SolidWorks

    (b) Inventor

    (c) Free CAD

    (d) All of the above

40. Which of the following is not a type of 3D printing technology?

    (a) Selective laser sintering

    (b) Forging

    (c) Fused deposition modelling

    (d) Stereolithography

41. What is the primary function of hearth use in the smithy shop?

    (a) Cool the object to desired temperature

    (b) Allow the air supply into the furnace

    (c) Heat the object to the desired temperature

    (d) Allow the passage for flue gases

42. The main operations done on a swage block are.

    (a) Sizing

    (b) Squaring

(c) Bending

(d) All of the above

43. Identify the electric switch shown in the figure.

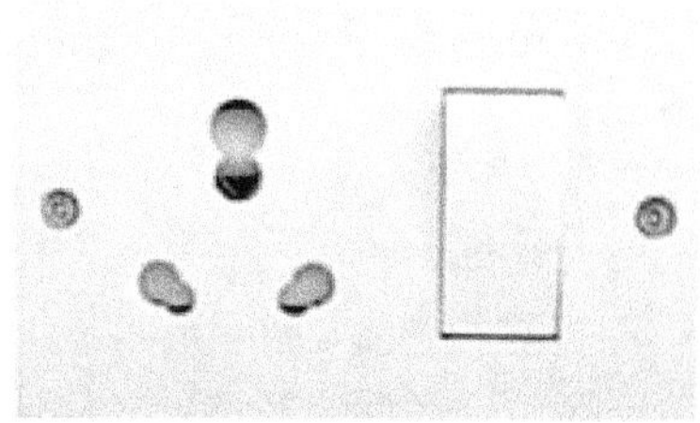

    (a) One way switch

    (b) Bell switch

    (c) Two-way switch

    (d) Power switch

44. Identify the tool used to uncover the insulation covering from electric cables.

    (a) Soldering iron

    (b) Wire stripper

    (c) Voltage tester

    (d) Insulated tape

45. Which of the following is not a lathe operation?

    (a) Punching

    (b) Knurling

    (c) Facing

    (d) Turning

46. What is the purpose of ball peen hammer?

    (a) Heavy duty work

    (b) Stretching the workpiece

    (c) Riveting and chipping

    (d) Bending

47. Which of the following is not a material used for a FDM 3D printer?
   (a) ABS (acrylonitrile butadiene styrene)
   (b) PLA (polylactic acid)
   (c) Steel
   (d) Aluminium

48. Identify the power tool as shown in figure.

   (a) Bench grinder
   (b) Heat gun
   (c) Hand drill
   (d) Angle grinder

49. Which of the following is a slicing software 3D printing?
   (a) Cura
   (b) PrusaSlicer
   (c) IdeaMaker
   (d) All of the above

50. Match the following steps of 3D printing as per their sequence.

	(a) Preparing the CAD model part	(b) Slicing and tool part geometry	(c) Convert CAD file to .STL file format	(d) 3D Printing of part
Step 1	○	○	○	○
Step 2	○	○	○	○
Step 3	○	○	○	○
Step 4	○	○	○	○

Note: Dark the circle in each column

51. Choose the following welding joints as per their applications.

	(a) Joining the edges of two plates kept in same plane	(b) Joining two overlapping plates	(c) Welding two workpieces at right angle	(d) Welding two parallel plates edge to edge
1. Tee joint	○	○	○	○
2. Edge joint	○	○	○	○
3. Lap joint	○	○	○	○
4. Butt joint	○	○	○	○

Note: Dark the circle in each column

52. Match the following as per their application.

	(a) Support working parts of lathe	(b) Support the main spindle	(c) Holds the cutting tool	(d) Controls tool movements in transverse and longitudinal direction
1. Carriage	◯	◯	◯	◯
2. Lathe bed	◯	◯	◯	◯
3. Tailstock	◯	◯	◯	◯
4. Headstock	◯	◯	◯	◯

53. Match the following as per their application.

	(a) Uniform packing of moulding sand	(b) Remove excess sand	(c) Clean the sand to remove unwanted material	(d) Making a hole in the cope
1. Rammer	◯	◯	◯	◯
2. Hand riddle	◯	◯	◯	◯
3. Sprue pin	◯	◯	◯	◯
4. Strike off bar	◯	◯	◯	◯

Note: Dark the circle in each column

Answer key

Q No.	Answer	Q No.	Answer	Q No.	Answer	Q No.	Answer
1	a	15	b	29	b	43	d
2	d	16	d	30	d	44	b
3	b	17	d	31	d	45	a
4	a	18	d	32	c	46	c
5	d	19	a	33	c	47	d
6	b	20	d	34	b	48	a
7	d	21	a	35	ab-bc-cd-da	49	a
8	a	22	d	36	d	50	1a-2c-3d-4b
9	d	23	a	37	a	51	1c-2d-3b-4a
10	d	24	a	38	d	52	1d-2a-3c-4b
11	d	25	b	39	d	53	1a-2c-3d-4b
12	d	26	b	40	b		
13	d	27	c	41	c		
14	d	28	d	42	d